Gilberto Bini - Corrado de Concini
Marzia Polito - Claudio Procesi

On the Work of Givental Relative to
Mirror Symmetry

APPUNTI

SCUOLA NORMALE SUPERIORE
1998

ISBN: 978-88-7642-240-9

ON THE WORK OF GIVENTAL RELATIVE TO MIRROR SYMMETRY

G. BINI, C. DE CONCINI, M. POLITO, AND C. PROCESI

ABSTRACT. These are the informal notes of two seminars held at the Università di Roma "La Sapienza", and at the Scuola Normale Superiore in Pisa in Spring and Autumn 1997.

We discuss in detail the content of the parts of Givental's paper [G1] dealing with mirror symmetry for projective complete intersections.

CONTENTS

Part 1. Introduction

Let X be a generic quintic threefold in $\mathbb{P}^4$. In 1991 Candelas, de la Ossa, Green and Parkes [COGP] "predicted" the numbers n_d of degree d rational curves in X conjecturing that the generating function

$$K(q) = 5 + \sum_{d=1}^{\infty} n_d d^3 \frac{q^d}{1-q^d} \tag{1.1}$$

could be recovered via elementary transformations from the hypergeometric series

$$\sum_{d=0}^{\infty} \frac{(5d)!}{(d!)^5} q^d,$$

which is annihilated by the Picard-Fuchs linear differential operator

$$D = \left(\frac{d}{dt}\right)^4 - 5\exp(t) \prod_{m=1}^{4} \left(5\frac{d}{dt} + m\right). \tag{1.2}$$

Indeed, pick a basis $\{f_i(t)\}_{i=0,...,3}$ of the space of solutions of $D(f) = 0$, introduce the new variable $T(t) = \frac{f_1(t)}{f_0(t)}$, and consider functions $g_i(T) := \frac{f_i(t(T))}{f_0(t(T))}$, $i = 0,...,3$. These functions form a basis of solutions of the following differential equation:

$$\left(\frac{d}{dt}\right)^2 \frac{1}{K(\exp t)} \left(\frac{d}{dt}\right)^2 g(t) = 0,$$

where $K(q)$ is the function in 1.1.

This conjecture was motivated by a fascinating phenomenon which is known by physicists as "mirror symmetry". By this they mean that given X as above, it is possible to construct a one parameter family Y_t of Calabi-Yau threefolds which are *mirrors* of the quintic X, i.e. the Hodge numbers of each Y_t enjoy the property $h^{r,s}(Y_t) = h^{3-r,s}(X)$. In this case, Y_t may be described as follows. Take the family of quintics

$$Y'_\lambda = \left\{ [y_0 \cdot ... \cdot y_4] \in \mathbb{P}^5 : y_0 \cdot ... \cdot y_4 = (\lambda)^{\frac{1}{5}} \left(y_0^5 + ... + y_4^5\right) \right\},$$

act by multiplications of variables by fifth roots of unity, and resolve the singularities to obtain

$$Y_t = \left\{ (u_0, ..., u_4) \in \mathbb{C}^5 : u_0 \cdot ... \cdot u_4 = \exp(t), u_0 + ... + u_4 = 1 \right\}.$$

So in this context, there seems to arise a relation between enumerative geometry in X and solutions of differential equations for periods of the family $\{Y_t\}$.

It is also possible to extend observations made for a generic quintic to a broad class of complete intersections varieties in projective spaces and to toric varieties. In the sequel, we shall assume that X is a smooth projective complete intersection of codimension r and degree $l = (l_1, ..., l_r)$ in $\mathbb{P}^n$, with $\sum_{i=1}^r l_i \leq n+1$. As for the quintic threefold, we consider a linear differential operator $D_\hbar$ of order $n + 1 - r$ which depends on a parameter $\hbar$. We still refer to $D_\hbar$ as the Picard-Fuchs operator associated with X. The explicit form of this operator is:

$$D_\hbar = \left(\hbar\frac{d}{dt}\right)^{n+1-r} - \exp(t)\prod_{j=1}^{r} l_j \prod_{m=1}^{l_j-1}\left(l_j\hbar\frac{d}{dt} + m\hbar\right).$$

Note that for $n = 4$ and $r = 1$, $D_\hbar$ coincides with the operator 1.2 up to the factor $\hbar^4$. Eventually, in [G2], Givental explains how to construct a family Y_t of *mirrors* from solutions of the differential equation $D_\hbar(f(t, \hbar)) = 0$.

Even in this more general setting, if we pick solutions for the differential equation $D_\hbar(f(t, \hbar)) = 0$, and manipulate them in a suitable way (we shall give details later), then it is conjectured that we obtain functions whose coefficients contain numbers n_d which *virtually* count rational curves on X. In fact, under the assumptions that the number of degree d curves in X is finite, the numbers n_d coincides exactly with the number of these curves (cfr.[Ma]).

Givental 's goal is to *formulate in a proper way and prove the link between solutions of Picard-Fuchs equation $D_\hbar(f(t, \hbar)) = 0$ and numbers of rational curves on X* .

Let us briefly sketch Givental 's strategy. The numbers counting rational curves on X (i.e. the Gromov Witten invariants of X) are all contained in a function called *potential*, of the following form:

$$\phi(t_0, ..., t_m) = \sum_{n_0+...+n_m \geq 3}\frac{t_0^{n_0}}{n_0!}...\frac{t_m^{n_m}}{n_m!}\sum_{\beta \in H_2(X,\mathbb{Z})} I_\beta(n_0, ..., n_m).$$

We construct a family $\{\nabla_\hbar(s)\}$ of connections on the tangent bundle $TH^*(X)$ to the cohomology of X: if we fix a basis $\{T_0, ..., T_m\}$ of $H^*(X)$, and coordinates $\{t_0, ..., t_m\}$, then we can take $\{T_0, ..., T_m\}$ as a global frame for the trivial bundle $TH^*(X)$. The connections are defined as follows:

$$\nabla_\hbar(T_i) = -\frac{1}{\hbar}\sum_{k=0}^{m}\phi_{ijk}dt_j T_k,$$

i.e., the structure constants of the connections are the third derivatives of the potential.

Pick a basis of solutions of $\nabla_\hbar(s) = 0$, and consider their components along T_0; the claim is that these functions $\{s_i\}$, suitably "manipulated" (as in the case of the quintic) form a basis of solutions for the Picard-Fuchs equation.

Let us summarize:

$$
\begin{array}{ccc}
X & & X \\
\downarrow & & \downarrow \\
\text{Mirror family } \{Y_t\} & & \text{Potential (numbers of rational curves on } X) \\
\downarrow & & \downarrow \\
\text{Picard Fuchs operator } D_\hbar & & \text{Connections } \nabla_\hbar \\
\downarrow & & \downarrow \\
\text{Solutions of } D_\hbar f = 0 & \xleftrightarrow{\text{manipulating}} & \text{Solutions of } \nabla_\hbar g = 0
\end{array}
$$

An important remark is that $\nabla_\hbar (s) = 0$ is a system of first order differential equations, and a basis of solutions gives us complete information about the coefficients of the system, i.e. about the potential; hence the second column of the previous diagram can be also interpreted in the other way: by this we mean that from solutions of Picard-Fuchs equation of the mirror family, we can recover the numbers of rational curves on X. The diagram should better look like:

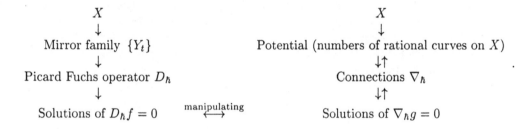

In writing these notes we followed very closely Givental's paper [G1]. For a different approach one can see [LLY]; also the recent Sem. Bourbaki talk [Pa] by R. Pandharipande contains a discusssion of these matters and further information on the literature. We are very grateful to him for reading a preliminary version of our notes and making many useful observations.

We heartily thank Enrico Arbarello for the extremely helpful conversations and suggestions during the development of our work.

Part 2. Preliminaries

2.1. QUANTUM COHOMOLOGY AND SMALL QUANTUM COHOMOLOGY

2.1.1. The case of convex varieties. Let X be a complex projective algebraic variety which satisfies the following properties:

- X is convex, i.e. for every morphism $\mu : \mathbb{P}^1 \to X$, $H^1(\mathbb{P}^1, \mu^*(T_X)) = 0$, with T_X the tangent bundle to X,
- $A^*(X) \cong H^{2*}(X, \mathbb{Z})$, i.e. the even integer cohomology is isomorphic to the Chow ring,
- the effective cone in $H_2(X, \mathbb{Z})$ is positively generated by a finite number of effective classes.

For the purpose of what follows, we may assume that X is a homogeneous space G/P, where G is a semisimple Lie group and P is a parabolic subgroup. Therefore the second hypothesis is automatically satisfied. Moreover, we shall deal with $\mathbb{P}^n$ for explicit calculations.

Under these hypotheses, we can build the moduli space of stable maps of n pointed genus 0 curves to X, whose image lies in the homology class $\beta \in H_2(X, \mathbb{Z})$, denoted by $\overline{\mathcal{M}}_{0,n}(X, \beta)$, which is an orbifold, and a locally normal projective variety of pure dimension $\dim X + \int_\beta c_1(TX) + n - 3$. Together with the moduli space we get the evaluation maps:

$$\rho_i : \quad \begin{matrix} \overline{\mathcal{M}}_{0,n}(X, \beta) & \to & X \\ [C, x_1, ..., x_n, \psi] & \to & \psi(x_i) \end{matrix} .$$

When we consider vector bundles on $\overline{\mathcal{M}}_{0,n}(X, \beta)$, we always assume them to be bundles in the *orbifold sense*, if not explicitely stated.

Let $\gamma_1, ..., \gamma_n$ be classes in $A^*(X)$, and $\beta \in H_2(X, \mathbb{Z})$; we define the Gromov Witten invariant:

$$I_\beta(\gamma_1, ..., \gamma_n) := \int_{\overline{\mathcal{M}}_{0,n}(X,\beta)} \rho_1^*(\gamma_1) \cup ... \cup \rho_n^*(\gamma_n).$$

Three properties of these invariants are needed:

1. $I_0(\gamma_1, ..., \gamma_n) = \begin{cases} \int_X \gamma_1 \cup \gamma_2 \cup \gamma_3 \text{ if } n = 3 \\ 0 \text{ if } n > 3 \end{cases}$

2. $I_\beta(1, \gamma_2, ..., \gamma_n) = \begin{cases} \int_X \gamma_2 \cup \gamma_3 \text{ if } n = 3 \text{ and } \beta = 0 \\ 0 \text{ otherwise} \end{cases}$

3. if $\gamma_1 \in A^1(X)$, then $I_\beta(\gamma_1, ..., \gamma_n) = \int_\beta \gamma_1 \cdot I_\beta(\gamma_2, ..., \gamma_n)$.

Let us fix a basis $\{T_0, T_1, ..., T_p, T_{p+1}, ..., T_m\}$ of $A^*(X)$ as a $\mathbb{Z}$−module, such that $T_0 = 1, < T_1, ..., T_p > = A^1(X)$. Consider the vector space $QH^*(X) = A^*(X) \otimes \mathbb{Q}[[t_0, ..., t_m]]$.
By means of the invariants, it is possible to define a product in this space so that it becomes the quantum cohomology ring of X.

Let $\gamma = t_0 T_0 + ... + t_m T_m$ be a generic element of $A^*(X) \otimes \mathbb{Q}[[t_0, ..., t_m]]$; the potential is defined as follows:

$$\phi(t_0, ..., t_m) \quad : \quad = \sum_{n \geq 3} \frac{1}{n!} \sum_{\beta \in H_2(X,\mathbb{Z})} I_\beta \left(\underbrace{\gamma, ..., \gamma}_{n} \right) =$$

$$= \sum_{n_0 + ... + n_m \geq 3} \frac{t_0^{n_0}}{n_0!} \cdots \frac{t_m^{n_m}}{n_m!} \sum_{\beta \in H_2(X,\mathbb{Z})} I_\beta \left(\underbrace{T_0, ..., T_0}_{n_0}, ..., \underbrace{T_m, ..., T_m}_{n_m} \right).$$

The third derivatives, which can be easily computed, have the following expression:

$$\phi_{ijk}(t_0, ..., t_m) = \frac{\partial^3 \phi}{\partial t_i \partial t_j \partial t_k} =$$

$$= \sum_{n \geq 0} \frac{1}{n!} \sum_{\beta \in H_2(X,\mathbb{Z})} I_\beta \left(T_i, T_j, T_k, \underbrace{\gamma, ..., \gamma}_{n} \right) =$$

$$= \sum_{n_0 + ... + n_m \geq 0} \frac{t_0^{n_0}}{n_0!} \cdots \frac{t_m^{n_m}}{n_m!}$$

$$\cdot \sum_{\beta \in H_2(X,\mathbb{Z})} I_\beta \left(T_i, T_j, T_k, \underbrace{T_0, ..., T_0}_{n_0}, ..., \underbrace{T_m, ..., T_m}_{n_m} \right).$$

If we denote by g^{ij} the inverse of the intersection matrix $g_{ij} = \int_X T_i \cup T_j$, we finally define the quantum product:

$$T_i * T_j := \sum_{k,l=0}^{m} \phi_{ijk} g^{kl} T_l.$$

Proposition 2.1. $(QH^*(X), *)$ is a commutative, associative algebra with unit T_0.

The next goal is to define the small quantum cohomology algebra, which, roughly speaking, involves only the potential restricted to the cohomology classes in $A^1(X)$; more precisely, we define the new product structure constants:

$$\psi_{ijk} := \phi_{ijk}(0, t_1, ..., t_p, 0, ..., 0).$$

Using properties of GW invariants, we see that

$$\phi_{ijk}(0, t_1, ..., t_p, 0, ..., 0) \quad =$$

$$= \sum_{n_1 + ... + n_p \geq 0} \frac{t_1^{n_1}}{n_1!} \cdots \frac{t_p^{n_p}}{n_p!} \cdot$$

$$\cdot \sum_{\beta \in H_2(X, \mathbb{Z})} I_\beta \left(T_i, T_j, T_k, \underbrace{T_1, ..., T_1}_{n_1}, ..., \underbrace{T_p, ..., T_p}_{n_p} \right) =$$

$$= \int_X T_i \cup T_j \cup T_k +$$

$$+ \sum_{\substack{\beta \in H_2(X, \mathbb{Z}) \\ \beta \neq 0}} \sum_{n_1 + ... + n_p \geq 0} \frac{t_1^{n_1}}{n_1!} \cdots \frac{t_p^{n_p}}{n_p!} \left(\int_\beta T_1 \right)^{n_1} \cdots \left(\int_\beta T_p \right)^{n_p} I_\beta(T_i, T_j, T_k) =$$

$$= \int_X T_i \cup T_j \cup T_k +$$

$$+ \sum_{\beta \neq 0} \left(\sum_{n_1 + ... + n_p \geq 0} \frac{t_1^{n_1}}{n_1!} \cdots \frac{t_p^{n_p}}{n_p!} \left(\int_\beta T_1 \right)^{n_1} \cdots \left(\int_\beta T_p \right)^{n_p} \right) I_\beta(T_i, T_j, T_k) =$$

$$= \int_X T_i \cup T_j \cup T_k + \sum_{\beta \neq 0} (\exp(t_1 \int_\beta T_1 + ... + t_p \int_\beta T_p)) I_\beta(T_i, T_j, T_k) =$$

$$= \int_X T_i \cup T_j \cup T_k + \sum_{\beta \neq 0} q_1^{\int_\beta T_1} \cdots q_p^{\int_\beta T_p} I_\beta(T_i, T_j, T_k).$$

where we have set $q_i = \exp(t_i)$, for $i = 1...p$.
Once more, setting

$$SQH^*(X) := A^*(X) \otimes \mathbb{Q}[[t_1, ..., t_p]]$$

and

$$T_i * T_j := \sum_{k,l=0}^{m} \psi_{ijk} g^{kl} T_l$$

one gets immediatly from Proposition 2.1

Corollary 2.2. $(SQH^*(X), *)$ *is a commutative, associative algebra with unit* T_0.

Although easier to deal with, the small quantum cohomology ring does not provide relevant information on the enumerative geometry of X.

Example 2.3.

$$SQH^*(\mathbb{P}^r) \cong \frac{\mathbb{Q}[p, q]}{(p^{r+1} - q)}.$$

Let us work out explicitly this example: first of all, we choose the basis of $H^*(\mathbb{P}^r) = \langle T_0, ..., T_r \rangle$, where $T_0 = 1, T_1 = p$ is the dual of the hyperplane class, and $T_i = p^i$. The product formula reduces to $T_i * T_j = \sum_{k=0}^m \psi_{ijk} T_{r-k}$.
About the computation of GW invariants, we observe that, taking the Poincaré dual α of p as a generator of $H_2(X, \mathbb{Z})$, the formula for ψ_{ijk} becomes:

$$\psi_{ijk} = \int_X T_i \cup T_j \cup T_k + \sum_{d \in \mathbb{N}} q^d I_d(T_i, T_j, T_k)$$

where the positive integer d means that we are considering the class αd.
Observe that $\dim \overline{\mathcal{M}}_{0,3}(\mathbb{P}^r, d\alpha) = r + \int_{d\alpha} c_i(T\mathbb{P}^r) + 3 - 3 = r + d(r+1)$; a necessary condition for $I_d(T_i, T_j, T_k) \neq 0$ is: $\deg(\rho_1^*(T_i) \cup \rho_2^*(T_j) \cup \rho_3^*(T_k)) = \dim \overline{\mathcal{M}}_{0,3}(\mathbb{P}^r, d\alpha)$, i.e. $i + j + k = r(d+1) + d$. This is possible only for $d = 1$, i.e. $i + j + k = 2r + 1$. We have to distinguish two cases:

1. $i + j \leq r$; in this case ψ_{ijk} has only the "classical" part:

$$\psi_{ijk} = \int_X T_i \cup T_j \cup T_k = \begin{cases} 1 \text{ if } k = r - i - j \\ 0 \text{ otherwise} \end{cases} .$$

2. $i + j > r$; in this case ψ_{ijk} has only the "quantum " part:

$$\psi_{ijk} = q I_1(T_i, T_j, T_k) = \begin{cases} q I_1(T_i, T_j, T_{2r+1-i-j}) \text{ if } k = 2r + 1 - i - j \\ 0 \text{ otherwise} \end{cases} .$$

It is sufficient to evaluate $I_1(T_i, T_j, T_{2r+1-i-j})$ for $i = 1, j = r$, since the associativity law brings all other information. With the enumerative interpretation, the invariant $I_1(T_1, T_r, T_r) = \int_\alpha T_1 \cdot I_1(T_r, T_r) = I_1(T_r, T_r)$ counts the lines on $\mathbb{P}^r$ passing through two generic point, and equals 1.
Finally, we can establish that $p^{r+1} = p * p^r = \psi_{1rr} = q$, and observe that this is the only relation in the small quantum cohomology algebra.

2.1.2. The case of complete intersections.
Let X be a smooth projective complete intersection, of codimension r and degree $l = (l_1, ..., l_r)$, $X \xrightarrow{i} \mathbb{P}^n$; generically, X does not satisfy our hypothesis of convexity, and up to now we cannot define its quantum cohomology ring; the next goal is to solve this problem.
Let $V := i^* (H^* (\mathbb{P}^n)) \subseteq A^*(X)$; V is a subspace of dimension $m+1 = n+1-r$ of the even cohomology of X; fix a basis $\{T_0, ..., T_m\}$ of V, with coordinates $\{t_0, ..., t_m\}$. We will define Gromov-Witten invariants on X only for classes in V, and then construct the quantum cohomology ring of X defining a product in $QH^*(X) := V \otimes \mathbb{Q}[[t_0, ..., t_m]]$.

Consider the vector bundle $W_{k,d}$ on $\overline{\mathcal{M}}_{0,k}(\mathbb{P}^n, d)$ whose fibers are

$$W_{k,d}\left([C, x_1, ..., x_k, \psi]\right) := H^0\left(C, \psi^*\left(\oplus_{i=1}^r \mathcal{O}\left(l_i\right)\right)\right),$$

and let $\mathcal{E}_{k,d} := Euler\left(W_{k,d}\right)$.

Now we are ready to define the GW invariants for X; let $(\gamma_1, ..., \gamma_k) \in V^k$, and choose $(\delta_1, ..., \delta_k) \in (H^*(\mathbb{P}^n))^k$ such that $\gamma_1 = i^*(\delta_1), ..., \gamma_k = i^*(\delta_k)$; fix $\beta \in H_2(X)$ such that $i_*(\beta) = d\alpha$, then:

Definition 2.4.

$$I_d(\gamma_1, ..., \gamma_k) := \int_{\overline{\mathcal{M}}_{0,k}(\mathbb{P}^n, d)} \rho_1^*(\delta_1) \cup ... \cup \rho_k^*(\delta_k) \cup \mathcal{E}_{k,d}.$$

We now prove that this is well defined, i.e. that $I_d(\gamma_1, ..., \gamma_k)$ does not depend on the choice of $(\delta_1, ..., \delta_k)$; for this we need the following lemma.

Lemma 2.5. $Ker\, i^* \subset Ann(E)$, where $E := Euler\left(\oplus_{i=1}^r \mathcal{O}(l_i)\right)$.

Proof. For every $\alpha \in H^*(\mathbb{P}^n)$, we have that $i_*(i^*(\alpha)) = E \cdot \alpha$.
□

Take $\delta_1' = \delta_1 + \epsilon_1$, with $\epsilon_1 \in Ker\, i^*$. Then

$$\int_{\overline{\mathcal{M}}_{0,k}(\mathbb{P}^n, d)} \rho_1^*(\delta_1') \cup ... \cup \rho_k^*(\delta_k) \cup \mathcal{E}_{k,d} =$$

$$= \int_{\overline{\mathcal{M}}_{0,k}(\mathbb{P}^n, d)} \rho_1^*(\delta_1) \cup ... \cup \rho_k^*(\delta_k) \cup \mathcal{E}_{k,d} + \int_{\overline{\mathcal{M}}_{0,k}(\mathbb{P}^n, d)} \rho_1^*(\epsilon_1) \cup ... \cup \rho_k^*(\delta_k) \cup \mathcal{E}_{k,d}.$$

Look at the following diagram:

$$\begin{array}{ccc}
\overline{\mathcal{M}}_{0,k+1}(\mathbb{P}^n, d) & \xrightarrow{\tau_1} & \mathbb{P}^n \\
\pi \downarrow & \nearrow \rho_1 & \\
\overline{\mathcal{M}}_{0,k}(\mathbb{P}^n, d) & &
\end{array}$$

Since $\mathcal{E}_{k,d} = \pi_* \tau_1^*(E)$, we get

$$\int_{\overline{\mathcal{M}}_{0,k}(\mathbb{P}^n, d)} \rho_1^*(\epsilon_1) \cup ... \cup \rho_k^*(\delta_k) \cup \mathcal{E}_{k,d} =$$

$$= \int_{\overline{\mathcal{M}}_{0,k}(\mathbb{P}^n, d)} \rho_1^*(\epsilon_1) \cup ... \cup \rho_k^*(\delta_k) \cup \pi_* \tau_1^*(E) =$$

$$= \int_{\overline{\mathcal{M}}_{0,k+1}(\mathbb{P}^n, d)} \pi^*(\rho_1^*(\epsilon_1) \cup ... \cup \rho_k^*(\delta_k)) \cup \tau_1^*(E) =$$

$$= \int_{\overline{\mathcal{M}}_{0,k+1}(\mathbb{P}^n, d)} \pi^*(\rho_2^*(\delta_2) \cup ... \cup \rho_k^*(\delta_k)) \cup \tau_1^*(\epsilon_1 E) = 0.$$

We will now justify definition 2.4. Consider the subset of $\overline{\mathcal{M}}_{0,k}(\mathbb{P}^n, d)$, which consists of equivalence classes of curves mapping to X:

$$S_{0,k}(X, d) = \{[C, x_1, ..., x_k, \psi] \in \overline{\mathcal{M}}_{0,k}(\mathbb{P}^n, d) : \psi(C) \subset X\}.$$

If $S_{0,k}(X, d)$ were a closed projective subvariety, with at most finite quotient singularities, and if $i_* : H_2(X) \to H_2(\mathbb{P}^n)$ were injective, then we could take $S_{0,k}(X, d)$ as the moduli space $\overline{\mathcal{M}}_{0,k}(X, \beta)$, and we could claim that $\mathcal{E}_{k,d}$ is its fundamental class in $\overline{\mathcal{M}}_{0,k}(\mathbb{P}^n, d)$. In fact, consider the section $\sigma \in \oplus_{i=1}^r \mathcal{O}(l_i)$ whose locus of zeroes is X. From this, build a section Γ of $W_{k,d}$: $\Gamma([C, x_1, ..., x_k, \psi]) := \psi^*(\sigma)$; note that $S_{0,k}(X, d)$ is exactly the zero locus of Γ, and from this the claim would follow.

Unfortunately, we do not know anything about $S_{0,k}(X, d)$, and we have to take 2.4 just as a definition.

Now, we can extend all the previous results: let $\gamma = t_0 T_0 + ... + t_m T_m$ be a generic element of $V \otimes \mathbb{Q}[[t_0, ..., t_m]]$; we define the potential

$$\phi(t_0, ..., t_m) = \sum_{n_0 + ... + n_m \geq 3} \frac{t_0^{n_0}}{n_0!} \cdots \frac{t_m^{n_m}}{n_m!} \sum_{d \in i_*(H_2(X, \mathbb{Z}))} I_d \left(\underbrace{T_0, ..., T_0}_{n_0}, ..., \underbrace{T_m, ..., T_m}_{n_m} \right),$$

consider the third derivatives:

$$\phi_{ijk}(t_0, ..., t_m) = \sum_{n_0 + ... + n_m \geq 0} \frac{t_0^{n_0}}{n_0!} \cdots \frac{t_m^{n_m}}{n_m!} \cdot$$

$$\cdot \sum_{d \in i_*(H_2(X, \mathbb{Z}))} I_d \left(T_i, T_j, T_k, \underbrace{T_0, ..., T_0}_{n_o}, ..., \underbrace{T_m, ..., T_m}_{n_m} \right)$$

and the inverse g^{ij} of the intersection matrix $g_{ij} = \int_X T_i \cup T_j$, and then finally define the quantum product:

$$T_i * T_j := \sum_{k,l=0}^m \phi_{ijk} g^{kl} T_l.$$

Proposition 2.6. $(QH^*(X), *)$ is a commutative, associative algebra with unit T_0.

Proof. Commutativity is trivial, and the fact that T_0 is the unity follows from property 2. For the associativity, one can repeat exactly the proof in [FP], with a suitable modification of Lemma 15, p. 37 we now explain. With the same notation, $D(A_1, A_2, d_1, d_2)$ is the boundary divisor of $\overline{\mathcal{M}}_{0,k}(\mathbb{P}^n, d)$ isomorphic to $\overline{\mathcal{M}}_{0,A_1 \cup \{\cdot\}}(\mathbb{P}^n, d_1) \times_{\mathbb{P}^n} \overline{\mathcal{M}}_{0,A_2 \cup \{\cdot\}}(\mathbb{P}^n, d_2)$, i is the natural inclusion of $D(A_1, A_2, d_1, d_2)$ in $\overline{\mathcal{M}}_{0,A_1 \cup \{\cdot\}}(\mathbb{P}^n, d_1) \times \overline{\mathcal{M}}_{0,A_2 \cup \{\cdot\}}(\mathbb{P}^n, d_2)$, and j the embedding of $D(A_1, A_2, d_1, d_2)$ in $\overline{\mathcal{M}}_{0,k}(\mathbb{P}^n, d)$.

Let us now compute the restriction of the Euler class $\mathcal{E}_{k,d}$ to the image of j; if $\rho^i_{k_i+1}$ is the evaluation map on the last point of $\overline{\mathcal{M}}_{0,A_i \cup \{\cdot\}}(\mathbb{P}^n, d_i)$, we define W'_{k_i+1,d_i} as the kernel of the map

$$W_{k_i+1,d_i} \ \rightarrow \ \rho^{i*}_{k_i+1}\left(\oplus^r_{i=1}\mathcal{O}(l_i)\right)$$
$$\sigma \ \rightarrow \ \sigma\left(\psi\left(x_{k_i+1}\right)\right),$$

therefore $\mathcal{E}'_{k_i+1,d_i} := Euler(W'_{k_i+1,d_i})$ satisfies $\mathcal{E}_{k_i+1,d_i} = \mathcal{E}'_{k_i+1,d_i}\rho^{i*}_{k_i+1}(E)$. If ν denotes the evaluation map on the meeting point in

$$D(A_1,A_2,d_1,d_2) \simeq \overline{\mathcal{M}}_{0,A_1\cup\{\cdot\}}(\mathbb{P}^n,d_1) \times_{\mathbb{P}^n} \overline{\mathcal{M}}_{0,A_2\cup\{\cdot\}}(\mathbb{P}^n,d_2)$$

then we get the exact sequence:

$$0 \ \rightarrow \ W'_{k_1+1,d_1}\oplus W'_{k_2+1,d_2} \ \rightarrow \ \begin{matrix} W_{k,d} \\ [\psi^*(\sigma),\psi^*(\tau)] \end{matrix} \ \begin{matrix} \rightarrow \\ \rightarrow \end{matrix} \ \begin{matrix} \nu^*\left(\oplus_i\mathcal{O}\left(l_i\right)\right) \\ \sigma\left(x_{k_1+1}\right)=\tau\left(x_{k_2+1}\right) \end{matrix} \ \rightarrow \ 0$$

and $\mathcal{E}_{k,d} = \mathcal{E}'_{k_1+1,d_1}\mathcal{E}'_{k_2+1,d_2}\nu^*(E)$.

Fix a suitable basis for $H^*(\mathbb{P}^n)$; let $\{T_0,...,T_s\}$ be a set of independent elements such that $\{[T_0],...,[T_s]\}$ form an orthonormal basis in $\frac{H^*(\mathbb{P}^n)}{Ann(E)}$, hence $\int_X T_i \cup T_j E = \delta_{ij}$. Let us denote $T_i E$ with T'_i and complete $\{T_0,...,T_s\}$ and $\{T'_0,...,T'_s\}$ to orthonormal dual bases of $H^*(\mathbb{P}^n)$ $\{T_0,...,T_s,T_{s+1},...,T_n\}$ and $\{T'_0,...,T'_s,T'_{s+1},...,T'_n\}$, such that the class of the diagonal in $\mathbb{P}^n$ is

$$\Delta = \sum_{i=0}^n T_i \otimes T'_i.$$

Putting everything toghether, and denoting with $\mu = \left(\rho^1_{k_1+1},\rho^2_{k_2+1}\right)$ the product of the evaluation maps on the last marked points, we get

$$\int_{D(A_1,A_2,d_1,d_2)} \rho^*_1(\delta_1)...\rho^*_k(\delta_k)\,\mathcal{E}_{k,d} =$$

$$= \int_{D(A_1,A_2,d_1,d_2)} \rho^*_1(\delta_1)...\rho^*_k(\delta_k)\,\mathcal{E}'_{k_1+1,d_1}\mathcal{E}'_{k_2+1,d_2}\mu^*(E\cdot(1\otimes 1))\,\mu^*(\Delta) =$$

$$= \int_{D(A_1,A_2,d_1,d_2)} \rho^*_1(\delta_1)...\rho^*_k(\delta_k)\,\mathcal{E}'_{k_1+1,d_1}\mathcal{E}'_{k_2+1,d_2}\mu^*\left(E\sum_{i=0}^m T_i \otimes T'_i\right) =$$

$$= \int_{D(A_1,A_2,d_1,d_2)} \rho^*_1(\delta_1)...\rho^*_k(\delta_k)\,\mathcal{E}'_{k_1+1,d_1}\mathcal{E}'_{k_2+1,d_2}\mu^*\left((E\otimes E)\sum_{i=0}^m T_i \otimes T_i\right) =$$

$$= \sum_{i=1}^n \int_{\overline{\mathcal{M}}_{0,A_1\cup\{\cdot\}}(\mathbb{P}^n,d_1)} \rho^*_1(\delta_1)...\rho^*_{k_1}(\delta_k)\rho^1_{k_1+1}(T_i)\,\mathcal{E}_{k_1+1,d_1} \ \cdot$$

$$\cdot \int_{\overline{\mathcal{M}}_{0,A_2\cup\{\cdot\}}(\mathbb{P}^n,d_2)} \rho^*_{k_1+1}(\delta_{k_1+1})...\rho^*_k(\delta_k)\rho^2_{k_2+1}(T_i)\,\mathcal{E}_{k_2+1,d_2} \ \cdot$$

$\square$

In the same way we can define the small quantum cohomology ring: $SQH^*(X) :=$ $V \otimes \mathbb{Q}[q]$,

$$\psi_{ijk}(q) := \sum_{d \in \iota_*(H_2(X,\mathbb{Z}))} q^d I_d(T_i, T_j, T_k),$$

and $T_i * T_j := \sum_{k,l=0}^m \psi_{ijk} g^{kl} T_l$.

Proposition 2.7. $(SQH^*(X), *)$ *is a commutative, associative algebra with unit* T_0.

Example 2.8. *In the case of a quintic threefold X in $\mathbb{P}^4$, $SQH^*(X) = QH^*(X)$.*

This follows from a dimension computation: in fact, $\dim_{\mathbb{R}} \overline{\mathcal{M}}_{0,k}(\mathbb{P}^4, d) = 10d + 2k + 2$, and $\deg \mathcal{E}_{k,d} = 10d + 2$, hence the non zero GW involve only degree two classes.

2.2. EQUIVARIANT COHOMOLOGY FOR TORUS ACTIONS

Let X be a topological space and $G \cong (S^1)^k$ be a compact torus acting on X. Recall the following definition:

Definition 2.9. *The universal bundle for a compact Lie group G is the G-fibering $EG \to BG = EG/G$, where EG is a contractible topological space on which G acts freely, and BG is called classifying space.*

Remark 2.10. *The universal bundle satisfies the universal property that every G-principal bundle is obtained from it by pullback.*

In the case of the torus, a model for the universal bundle is the Hopf fibering of infinite dimensional spheres on infinite dimensional complex projective spaces:

$$\begin{array}{ccc} E(S^1)^k & \cong & (S^\infty)^k \\ \downarrow & & \\ B(S^1)^k & \cong & (\mathbb{P}^\infty)^k \end{array} .$$

The group G acts on $X \times EG$, and we denote by $X_G = X \times_G EG$ the quotient; it is a bundle over BG, with fiber X:

$$\begin{array}{cccc} \pi: & X_G & \to & BG \\ & [x, e] & & [e] \end{array} .$$

Definition 2.11. *The G-equivariant cohomology of X is the cohomology of X_G:*

$$H_G^*(X) := H^*(X_G).$$

From now on we will consider cohomology with coefficients in $\mathbb{Q}$.

Note that $H_G^*(\{pt\}) = H^*(BG)$. Therefore the equivariant cohomology ring is a module (via π^*) over this ring.

In our case, $H_{S^1}^*(\{pt\}) = H^*(B(S^1)) = H^*(\mathbb{P}^\infty) \cong \mathbb{Q}[\lambda]$, and $H_{(S^1)^k}^*(\{pt\}) \cong \mathbb{Q}[\lambda_1, ..., \lambda_k]$.

The following example, and others which will come later in this section, are the most relevant from our point of view, since Givental uses the equivariant theory almost exclusively in these cases.

Example 2.12. $X = \mathbb{P}^1$, $G = (S^1)$.

The action is:

$$\begin{array}{ccccc} G & \times & X & \rightarrow & X \\ t & , & [x_0, x_1] & & [t^2 x_0, x_1] \end{array} \quad . \tag{2.1}$$

The bundle $\mathbb{P}_{S^1}^1 = \mathbb{P}^1 \times_{S^1} S^\infty \rightarrow \mathbb{P}^\infty$ is nothing else but the projectivization of a two dimensional complex vector bundle on $\mathbb{P}^\infty$, namely $\mathbb{P}(\mathcal{O}(-2) \oplus \mathcal{O}) \rightarrow \mathbb{P}^\infty$. Assuming that the cohomology of $\mathbb{P}^\infty$ is generated by $c_1(\mathcal{O}(-2))$, by the Whitney formula, we can compute:

$$H_G^*(X) = H^*(\mathbb{P}(\mathcal{O}(-2) \oplus \mathcal{O})) \cong \frac{\mathbb{Q}[\lambda, p]}{(p^2 - p\lambda)},$$

where p is to be interpreted as the generator of the cohomology of the fiber of the projective bundle $\mathbb{P}(\mathcal{O}(-2) \oplus \mathcal{O}) \rightarrow \mathbb{P}^\infty$.

Example 2.13. *A generalization to a diagonal action of a k dimensional torus G on $\mathbb{P}^n$, with characters $\chi_i \in \mathbb{Z}^k$, $i = 0, ..., n$.*

Let L_j be the pull backs of the hyperplane bundle on the different components of $(\mathbb{P}^\infty)^k$, and take the Chern classes of these bundles as generators of the cohomology of $(\mathbb{P}^\infty)^k$; we have:

$$\mathbb{P}^n \times_G (S^\infty)^k \cong \mathbb{P}\left((\otimes_{j=1}^k L_j^{\chi_{oj}}) \oplus ... \oplus (\otimes_{j=1}^k L_j^{\chi_{nj}})\right) \rightarrow (\mathbb{P}^\infty)^k.$$

Setting $\chi_i(\lambda) := \prod_j \lambda_j^{\chi_{ij}}$, and applying Whitney formula,

$$H_G^*(\mathbb{P}^n) = \frac{\mathbb{Q}[\lambda_1, ..., \lambda_k, p]}{\prod_{i=0,...,n}(p - \chi_i(\lambda))}.$$

2.2.1. G-vector bundles and equivariant characteristic classes. Suppose we have a complex rank r vector bundle V over X with a G linear action, compatible with the projection onto X and linear on the fibers, then $V_G := V \times_G EG$ is still a $\mathbb{C}^r$ bundle over X_G. We define the equivariant Chern and Euler classes of V:

$$\begin{aligned} c_i^G(V) &:= c_i(V_G) \\ \mathcal{E}^G(V) &:= \mathcal{E}(V_G). \end{aligned}$$

If we reconsider Examples 2.12 and 2.13 more closely, we realize that the actions described there correspond to a distinguished choice of linearization of the tautological bundles over projective spaces. If another choice had been made, then equivariant cohomology groups would have been isomorphic. We now give a further example which will be useful for the purpose of what folllows.

Example 2.14. $X = \mathbb{P}^n$, $V = \oplus_{1=i}^r \mathcal{O}(l_i)$, $G = T^{n+1} \times T^r$. *We compute the equivariant Euler class.*

T^{n+1} acts on X as in case 2.13, with $k = n + 1$, and $\chi_i = (0, ..., 0, \underset{i}{1}, 0, ..., 0)$, and T^r acts trivially; therefore $H_T^*(\mathbb{P}^n) \cong \frac{\mathbb{Q}[\lambda_0,...,\lambda_n,\mu_1,...,\mu_r,p]}{\prod_{i=0}^n (p-\lambda_i)}$. Conversely, T^r acts on a fiber of the bundle by diagonal action with characters $\mu_i = (0, ..., 0, \underset{i}{1}, 0, ..., 0)$, and the T^{n+1}action is induced by the one on X.

Let's look at the following diagram:

$$\oplus_{i=1}^r \mathcal{O}(l_i) \times_{(S^1)^{n+1}} (S^\infty)^{n+1} \times_{(S^1)^r} (S^\infty)^r \overset{\phi}{\to} \mathbb{P}^n \times_{(S^1)^{n+1}} (S^\infty)^{n+1} \times (\mathbb{P}^\infty)^r$$
$$\overset{\pi'}{\searrow} \qquad \qquad \downarrow \pi$$
$$(\mathbb{P}^\infty)^{n+1} \times (\mathbb{P}^\infty)^r$$

Restricted to a fiber of π, which is isomorphic to $\mathbb{P}^n$, the diagram looks like:

$$\oplus_{i=1}^r \mathcal{O}(l_i) \otimes M_i^{-1} \overset{\phi}{\to} \mathbb{P}^n \times (\mathbb{P}^\infty)^r ,$$

where M_i is the pull-back of the hyperplane bundle on the i-th copy of $\mathbb{P}^\infty$, and has Chern class μ_i.

The Euler class of this restricted bundle is $\prod_{i=1}^r (l_i p - \mu_i)$; using this, we can prove that $\mathcal{E}^G(V) = \prod_{i=1}^r (l_i p - \mu_i)$.

2.2.2. Equivariant integral and pairing. Let now X be a compact manifold with at most finite quotient singularities.The fibering $\pi : X_G \to BG$ induces a push-forward map $\pi_* : H_G^*(X) \to H_G^*(\{pt\})$, which we will call equivariant integral and often denote with $\int^G$. All the same for the equivariant pairing

$$\langle,\rangle_G : \quad H_G^*(X) \times H_G^*(X) \to H_G^*(\{pt\})$$
$$(\omega, \eta) \qquad \qquad \pi_*(\omega \cup \eta).$$

Example 2.15. *Case 2.13.*

The equivariant integral is given by the following formula:

$$\int^G f(p, \lambda_0, ..., \lambda_n) = \frac{1}{2\pi\sqrt{-1}} \int \frac{f(p, \lambda_0, ..., \lambda_n)}{\prod_{j=0}^n (p - \chi_j(\lambda))} dp.$$

In fact, π_* is just the integration along the fiber, which is a n dimensional projective space whose cohomology is generated by p. Therefore, writing $f(p, \lambda_0, ..., \lambda_n) =$

$\sum_j f_j(\lambda) p^j$, we see that $\int^T f(p, \lambda_0, ..., \lambda_n) = f_n(\lambda)$. On the other hand,

$$\frac{1}{2\pi\sqrt{-1}} \int \frac{\sum_j f_j(\lambda) p^j}{\prod_j (p - \chi_j)} dp = \frac{1}{2\pi\sqrt{-1}} \sum_j f_j(\lambda) \int \frac{p^j}{\prod_j (p - \chi_j)} dp;$$

these integrals all vanish by degree computation except for $j = n$, and the formula follows.

2.2.3. Localization at fixed points and integration formula.

One of the main tools of equivariant cohomology is the localization at fixed points. Equivariant cohomology rings for torus actions are modules over $\mathbb{Q}[\lambda_1, ..., \lambda_k]$, and therefore it is possible to localize on their support. Suppose $D_1, ..., D_s$ are the connected components of fixed points for the T action on X, then $H_T^*(D_i) = H^*(D_i \times_T ET) = H^*(D_i \times BT) \cong H^*(D_i) \otimes H_T^*(\{pt\}) \cong H^*(D_i) \otimes \mathbb{Q}[\lambda_1, ..., \lambda_k]$.

Proposition 2.16. *There is a natural morphism*

$$H_T^*(X) \quad \overset{\delta=(\delta_1,...,\delta_s)}{\to} \quad \oplus_{i=1...,s} H_T^*(D_i) \ ,$$

whose kernel and cokernel are torsion modules.

Example 2.17. *Case 2.12.*

There are two fixed points, $[0, 1]$, and $[1, 0]$ (say 0 and ∞). The morphism is

$$
\begin{array}{ccc}
H_T^*(\mathbb{P}^1) & \to & H_T^*(\{0\}) \oplus H_T^*(\{\infty\}) \\
\cong & & \cong \\
\frac{\mathbb{Q}[\lambda,p]}{(p^2 - p\lambda)} & \to & \mathbb{Q}[\lambda] \oplus \mathbb{Q}[\lambda] \\
f(p, \lambda) & & (f(\lambda, \lambda), f(0, \lambda))
\end{array}
$$

Localizing both sides outside the ideal $\langle \lambda \rangle$, we get an isomorphism. In particular, we see that

$$
\begin{array}{ccc}
\frac{p}{\lambda} & \to & (1, 0) \\
\frac{\lambda - p}{\lambda} & \to & (0, 1)
\end{array}
$$

and that $1 = \frac{p}{\lambda} + \frac{\lambda - p}{\lambda}$.

Example 2.18. *Case 2.13, with $k = n + 1$, and $\chi_i = (0, ..., 0, \underset{i}{1}, 0, ..., 0)$, as in 2.2.1.*

There are $n + 1$ fixed points, $x_i = \left[0, ..., 0, \underset{i}{1}, 0, ..., 0\right]$; the morphism is

$$
\begin{array}{ccc}
\frac{\mathbb{Q}[\lambda_0, ..., \lambda_n, p]}{\prod_{i=0}^n (p - \lambda_i)} & \to & \oplus_{i=0,...,n} \mathbb{Q}[\lambda_0, ..., \lambda_n] \\
f(p, \lambda_0, ..., \lambda_n) & & f(\lambda_i, \lambda_0, ..., \lambda_n)
\end{array}
$$

The isomorphism comes localizing both sides ouside the ideal $\langle \lambda_0, ..., \lambda_n \rangle$. Once more, it is useful to know that, if we set $\phi_i = \frac{\prod_{j \neq i}(p - \lambda_j)}{\prod_{j \neq i}(\lambda_i - \lambda_j)}$, then ϕ_i goes to $(0, ..., 0, \underset{i}{1}, 0, ..., 0)$.

Notice that after having localized,

$$H^*_G(X)_{\langle \lambda_0, ..., \lambda_n \rangle} \cong \frac{\mathbb{Q}(\lambda_0, ..., \lambda_n)\,[p]}{\prod_{i=0}^{n}(p - \lambda_i)}$$

becomes a vector space of dimension $n + 1$ over the field $\mathbb{Q}(\lambda_0, ..., \lambda_n)$, and that the equivariant pairing gives a scalar product. The ϕ_i's form an orthogonal basis with respect to this product: if $i \neq k$, $\phi_i \cdot \phi_k$ vanishes, and on the other hand

$$\int^G \phi_i^2 = \frac{1}{2\pi\sqrt{-1}\prod_{j \neq i}(\lambda_i - \lambda_j)^2} \int^G \frac{\prod_{j \neq i}(p - \lambda_j)^2}{\prod_j (p - \lambda_j)} dp =$$

$$= \frac{1}{2\pi\sqrt{-1}\prod_{j \neq i}(\lambda_i - \lambda_j)^2} \int^G \frac{\prod_{j \neq i}(p - \lambda_j)}{(p - \lambda_i)} dp = \frac{1}{\prod_{j \neq i}(\lambda_i - \lambda_j)}$$

The integration formula gives us a method for reducing equivariant integrals to integrals on fixed points components; since equivariant cohomology on fixed points is just usual cohomology tensorized by a ring of polynomials, the formula simplifies computations in the sense that it reduces the problem of integrating along the fiber isomorphic to X to the one of integrating on some subvarieties of X; let $\mathcal{E}_i$ be the equivariant Euler class of the normal bundle $\mathcal{N}_{D_i/X}$ of the connected component of fixed points D_i in X.

Proposition 2.19. *The following formula holds:*

$$\int^G_X \omega = \sum_{i=1}^{s} \int^G_{D_i} \frac{\delta_i(\omega)}{\mathcal{E}_i}.$$

Example 2.20. *Case 2.12.*

The normal bundle of $\{0\}$ in $\mathbb{P}^1$ can be identified equivariantly with $\mathbb{P}^1 \backslash \{\infty\} = \mathbb{C}$, and S^1 acts on it by multiplication by t^2; the Euler class of this bundle is λ . Similarly, the action of S^1 on the normal bundle to $\{\infty\}$ is by multiplication by t^{-2} and the Euler class is $-\lambda$. The formula is then:

$$\int^{S^1}_{\mathbb{P}^1} f(p, \lambda) = \frac{f(\lambda, \lambda)}{\lambda} - \frac{f(0, \lambda)}{\lambda}.$$

Example 2.21. $G = S^1$ *acts on* $X \times \mathbb{P}^1$, *trivially on the first factor, and as in 2.12 on the second.*

This example does not bring anything new from the point of view of equivariant cohomology, but is frequently used by Givental.

Clearly, $H^*_G(X \times \mathbb{P}^1) \cong H^*(X) \otimes H^*_G(\mathbb{P}^1)$, and the localization isomorphism tells us:

$$H^*(X) \otimes \frac{\mathbb{Q}(\lambda)\,[p]}{(p^2 - p\lambda)} \cong (H^*(X) \otimes \mathbb{Q}(\lambda)) \oplus (H^*(X) \otimes \mathbb{Q}(\lambda)).$$

If $x \in H^*_G(X \times \mathbb{P}^1)$ and $\delta(x) = (t, \tau)$, the same as giving the decomposition $x = t\frac{p}{\lambda} + \tau\frac{\lambda - p}{\lambda}$, we will say that x is of type 0 (resp. ∞) if $\tau = 0$ (resp. $t = 0$). The equivariant integral is given by

$$\int_{X \times \mathbb{P}^1}^{S^1} x = \frac{\int_X t - \int_X \tau}{\lambda},$$

and the equivariant pairing is written as follows:

$$\langle x, x' \rangle_{X \times \mathbb{P}^1}^G = \frac{\langle t, t' \rangle_X - \langle \tau, \tau' \rangle_X}{\lambda}.$$

2.3. EQUIVARIANT GROMOW WITTEN THEORY

Keeping the notations of the previous sections, a T action on X induces a T action on the space of maps of curves in X, as it acts on the image of every map. Since the action is compatible with the equivalence relation on maps, we define an induced action on $\overline{\mathcal{M}}_{0,n}(X, \beta)$:

$$\begin{array}{ccc}
T \times \overline{\mathcal{M}}_{0,n}(X, \beta) & \to & \overline{\mathcal{M}}_{0,n}(X, \beta) \\
(t, [C, x_1, ..., x_n, \psi]) & & [C, x_1, ..., x_n, t \circ \psi]
\end{array}.$$

The evaluation maps are equivariant, and therefore induce morphisms in equivariant cohomolgy:

$$\rho_i^* : H_T^*(X) \to H_T^*(\overline{\mathcal{M}}_{0,n}(X, \beta)).$$

Fix a basis $\{1 = T_0, ..., T_m\}$ of $H_T^*(X)$ as $H_T^*(\{pt\})-$ module, and take n classes $\gamma_1, ..., \gamma_n$ in it, and a class $\beta \in H_2(X)$.

Definition 2.22. *The equivariant Gromov Witten invariants are:*

$$I_\beta^T(\gamma_1, ..., \gamma_n) := \int_{\overline{\mathcal{M}}_{0,n}}^T \rho_1^*(\gamma_1) \cup ... \cup \rho_n^*(\gamma_n).$$

They satisfy the three properties listed in 2.1, with the usual integral replaced by the equivariant one.

Now it is possible to repeat all constructions of the non equivariant case, i.e. define the potential

$$\mathcal{F} = \sum_{n \geq 3} \frac{1}{n!} \sum_{\beta \in H_2(X, \mathbb{Z})} I_\beta^T \left(\underbrace{\gamma, ..., \gamma}_{n} \right),$$

evaluate its third derivatives

$$\mathcal{F}_{ijk} = \frac{\partial^3 \mathcal{F}}{\partial t_i \partial t_j \partial t_k} = \sum_{n \geq 0} \frac{1}{n!} \sum_{\beta \in H_2(X,\mathbb{Z})} I_\beta^T \left(T_i, T_j, T_k, \underbrace{\gamma, ..., \gamma}_{n} \right),$$

consider the vector space $QH_T^*(X) = H_T^*(X) \otimes \mathbb{Q}[[t_0, ..., t_m]]$, and define on it the equivariant quantum product

$$T_i * T_j = \sum_{k,l=0}^{m} \mathcal{F}_{ijk} g^{kl} T_l,$$

where g^{ij} is still the inverse of the intersection matrix with respect to the chosen basis of $H_T^*(X)$.

Proposition 2.23. $(QH_T^*(X), *)$ *is a commutative, associative algebra with unit* T_0.

Example 2.24. *Compute* $QH_{S^1}^*(\mathbb{P}^2)$, *with the action:*

$$\begin{array}{ccccc}
S^1 & \times & \mathbb{P}^2 & \to & \mathbb{P}^2 \\
t & , & [x_0 : x_1 : x_2] & & [t^a x_0 : t^b x_1 : t^c x_2]
\end{array}.$$

As a first step, we compute the equivariant cohomology $H_{S^1}^*(\mathbb{P}^2)$; the bundle $\mathbb{P}^2 \times_{s^1} \mathbb{P}^\infty \to \mathbb{P}^\infty$ is exactly the bundle $\mathbb{P}(\mathcal{O}(-a) \oplus \mathcal{O}(-b) \oplus \mathcal{O}(-c)) \to \mathbb{P}^\infty$; following 2.13,

$$H_{S^1}^*(\mathbb{P}^2) \cong \frac{\mathbb{Q}[\lambda, p]}{(p - a\lambda)(p - b\lambda)(p - c\lambda)} = \frac{\mathbb{Q}[\lambda, p]}{(p^3 - e_1 \lambda p^2 + e_2 \lambda^2 p - e_3 \lambda^3)},$$

where we have set: $e_1 = a + b + c$, $e_2 = ab + ac + bc$, $e_3 = abc$.
A basis for $H_{S^1}^*(\mathbb{P}^2)$ as a $\mathbb{Q}[\lambda]$ - module is $\{T_0, T_1, T_2\}$, with $T_i = p^i$. Let us compute $\int_{\mathbb{P}^2}^{S^1} T_i \cup T_j = \int_{\mathbb{P}^2}^{S^1} p^{i+j}$.
Since $i + j \leq 4$, it is sufficient to compute:

$$\int_{\mathbb{P}^2}^{S^1} p^0 = \int_{\mathbb{P}^2}^{S^1} p^1 = 0,$$

$$\int_{\mathbb{P}^2}^{S^1} p^2 = 1,$$

$$\int_{\mathbb{P}^2}^{S^1} p^3 = \int_{\mathbb{P}^2}^{S^1} e_1 \lambda p^2 - e_2 \lambda^2 p + e_3 \lambda^3 = e_1 \lambda,$$

$$\int_{\mathbb{P}^2}^{S^1} p^4 = \int_{\mathbb{P}^2}^{S^1} (e_1^2 - e_2) \lambda^2 p^2 - (e_1 e_2 - e_3) \lambda^3 p + e_1^2 \lambda^4 = (e_1^2 - e_2) \lambda^2,$$

in order to write down the intersection matrix and its inverse:

$$g_{ij} = \begin{pmatrix} 0 & 0 & 1 \\ 0 & 1 & e_1\lambda \\ 1 & e_1\lambda & (e_1^2 - e_2)\lambda^2 \end{pmatrix}, g^{ij} = \begin{pmatrix} e_2\lambda^2 & -e_1\lambda & 1 \\ -e_1\lambda & 1 & 0 \\ 1 & 0 & 0 \end{pmatrix}.$$

Note that $\dim \overline{\mathcal{M}}_{0,k}\left(\mathbb{P}^2, d\right) = 2 + \int_d c_1(T\mathbb{P}^2) + k - 3 = 3d + k - 1$, and, by degree computation and properties of GW invariants, using the notation $I_d(T_0^{n_0}, T_1^{n_1}, T_2^{n_2})$ for equivariant invariants, we see that:

1. if $n_0 > 0$, the GW invariants that do not vanish are just $I_0(T_0 \cup T_1 \cup T_1)$ and $I_0(T_0 \cup T_0 \cup T_2)$, and they are both equal to 1;

2. if $n_0 = 0$, $I_d(T_1^{n_1}, T_2^{n_2}) = \left(\int_d^T T_1\right)^{n_1} I_d(T_2^{n_2}) = d^{n_1} I_d(T_2^{n_2})$; since equivariant integral on moduli space is a push forward along a fiber of complex dimension $\geq 3d - 1$, $I_d(T_2^{n_2}) = 0$ for $n_2 < 3d - 1$.

The potential is:

$$
\begin{aligned}
\mathcal{F} &= \sum_{n_0+n_1+n_2 \geq 3} \frac{y_0^{n_0} \, y_1^{n_1} \, y_2^{n_2}}{n_0! \, n_1! \, n_2!} \sum_{d \geq 0} I_d \left(T_0^{n_0}, T_1^{n_1}, T_2^{n_2}\right) = \\
&= \mathcal{F}_{cl} + \sum_{d>0} \sum_{n_1 \geq 0} \sum_{k \geq 0} \frac{d^{n_1} y_1^{n_1}}{n_1!} \frac{y_2^{3d-1+k}}{(3d-1+k)!} I_d \left(T_2^{3d-1+k}\right) = \\
&= \mathcal{F}_{cl} + \sum_{d>0} \exp(dy_1) \sum_{k \geq 0} \frac{y_2^{3d-1+k}}{(3d-1+k)!} I_d \left(T_2^{3d-1+k}\right),
\end{aligned}
$$

Notice that, since $\dim_{\mathbb{R}} \overline{\mathcal{M}}_{0,3d-k+1}\left(\mathbb{P}^2, d\right) = 2(3d-1+3d-1+k) = 2(6d-2+k)$, and T_2^{3d-1+k} is a form of degree $4(3d-1+k)$, the push forward $I_d\left(T_2^{3d-1+k}\right)$ has degree $2k = 12d - 4 + 4k - 12d + 4 - 2k$. This implies, looking at $H_{s_1}^*(\{pt\})$, that $I_d\left(T_2^{3d-1+k}\right) = N_{d,k} \lambda^k$.

Working out calculations, as in [Al] , the associativity equation yields:

$$
\mathcal{F}_{222} + \mathcal{F}_{111}\mathcal{F}_{122} - \mathcal{F}_{112}^2 = -\left(e_2 + e_1^2\right) \lambda^2 \mathcal{F}_{112} + 2e_1 \lambda \mathcal{F}_{122} + \left(e_1 e_2 - e_3\right) \lambda^3 \mathcal{F}_{111}.
$$

From this we get recursive relations for $N_{d,k}'s$:

$$
\begin{aligned}
N_{d,k} = &\sum_{\substack{d_1+d_2=d\geq 1 \\ k_1+k_2=k\geq 0}} N_{d_1,k_1} N_{d_1,k_1} d_1 d_2 \left[d_1 d_2 \binom{3d-4+k}{3d_1-2+k_1} - d_1^2 \binom{3d-4+k}{3d_1-1+k_1} \right] + \\
&+ 2e_1 d N_{d,k-1} - (e_1^2 + e_2)d^2 N_{d,k-2} + (e_1 e_2 - e_3) d^3 N_{d,k-3}.
\end{aligned}
$$

This formula is a deformation of Kontsevich's one in the non-equivariant setting (see [FP]).

Part 3. Numbers counting curves and differential equations

3.1. CONNECTIONS

Let V be the vector subspace of $H^*(X)$ defined in 2.1. V is equipped with the non-degenerate bilinear form of intersection pairing; we choose an orthonormal basis $\{T_0, ..., T_m\}$ and coordinates $\{t_0, ..., t_m\}$, and recall that we have defined the potential ϕ and the quantum product on $QH^*(X) = V \otimes \mathbb{Q}[[t_0, ..., t_m]]$, namely $T_i * T_j = \sum_{k=0}^m \phi_{ijk} T_k$.

The tangent bundle TV is obviously trivial, and we can take $\{T_0, ..., T_m\}$ as a global frame; we give the formal definition of a family of connections on TV:

Definition 3.1.

$$\nabla_\hbar := d - \frac{1}{\hbar} \sum_{i=0}^m dt_i T_i *$$

Remark 3.2. *Since* $\nabla_\hbar(T_j) = -\frac{1}{\hbar} \sum_{i,k=0}^m \phi_{ikj} dt_i T_k$, *the structure constants of the connections are exactly the third derivatives of the potential, divided by* $\hbar$.

Proposition 3.3. $\nabla_\hbar$ *is flat.*

Proof. Flatness of the connection is equivalent to associativity of the quantum product. In fact, the matrix of 1-forms associated with the connection is $\Omega_{jk}^\hbar = \frac{1}{\hbar} \sum_{i=0}^m \phi_{ikj} dt_i = \frac{1}{\hbar} d(\phi_{kj})$; the matrix of the curvature $\nabla_\hbar^2$ is $K^\hbar = d\Omega^\hbar - \Omega^\hbar \cup \Omega^\hbar = -\Omega^\hbar \cup \Omega^\hbar$; since

$$
\begin{aligned}
K_{jk}^\hbar &= -\frac{1}{\hbar^2} \sum_{i=0}^m d(\phi_{ji}) \cup d(\phi_{ik}) = -\frac{1}{\hbar^2} \sum_{i=0}^m \left(\left(\sum_{p=0}^m \phi_{pij} dt_p \right) \cup \left(\sum_{q=0}^m \phi_{qik} dt_q \right) \right) = \\
&= -\frac{1}{\hbar^2} \sum_{i=0}^m \left(\left(\sum_{p=0}^m \phi_{pij} dt_p \right) \cup \left(\sum_{q=0}^m \phi_{qik} dt_q \right) \right) = \\
&= -\frac{1}{\hbar^2} \sum_{p,q=0}^m \left(\sum_{i=0}^m \phi_{pij} \phi_{qik} \right) dt_p \cup dt_q = \\
&= \sum_{\substack{p,q=0 \\ p<q}}^m \left(-\frac{1}{\hbar^2} \sum_{i=0}^m (\phi_{pij} \phi_{qik} - \phi_{qij} \phi_{pik}) \right) dt_p \cup dt_q,
\end{aligned}
$$

then

$$K^\hbar = 0 \iff \sum_{i=0}^m (\phi_{pij} \phi_{qik} - \phi_{qij} \phi_{pik}) \quad \forall p, q, j, k.$$

On the other side,

$$(T_j * T_p) * T_k - T_j * (T_p * T_k) \;=\; \sum_i \phi_{jpi} \, (T_i * T_k) - \sum_i \phi_{kpi} \, (T_j * T_i) =$$

$$= \sum_{q=0}^{m} \left(\sum_{i=0}^{m} (\phi_{jpi}\phi_{ikq} - \phi_{kpi}\phi_{jiq}) \right) T_q,$$

and the quantum product is associative if and only if $\sum_{i=0}^{m} (\phi_{pij}\phi_{qik} - \phi_{qij}\phi_{pik})$ $\forall p, q, j, k{=}0$.

$\square$

Corollary 3.4. *For every $\hbar$, the differential equation $\nabla_\hbar s(t, \hbar) = 0$ formally admits $m + 1$ linearly independent solutions.*

A remarkable construction of the solutions will be given soon using equivariant quantum cohomology. However, let us show first how we can recover the coefficients of the differential equation, i.e the derivatives of the potential, once we have a basis of independent solutions.
A vector field $s(t, \hbar) = \sum_{i=0}^{m} s_i(t, \hbar) T_i$ is a solution if and only if it satisfies:

$$\hbar \frac{\partial s_j}{\partial t_k}(t, \hbar) \;=\; \sum_{i=0}^{m} s_i \, (t, \hbar) \, \phi_{ijk}(t)$$

$$\forall j, k \;=\; 0, ..., m.$$

Conversely, given a basis $\{s_\beta(t, \hbar) = \sum_{i=0}^{m} s_{\beta i}(t, \hbar) T_i\}_{\beta=0,...,m}$ of independent solutions, then the $\phi_{ijk}(t)$ can be easily recovered with linear algebra computations.

Every time we deal with differential equations, we skip problems of existence, that is, our solutions are formal solutions; typically the existence in a formal neighborhood of the initial conditions is ensured. For details about the analytical point of view, see [Du].

3.2. EQUIVARIANT POTENTIAL AND SOLUTIONS OF $\nabla_\hbar s = 0$

We consider $Y := X \times \mathbb{P}^1$ with the S^1 action trivial on X as in 2.21:

$$\begin{array}{ccccc} S^1 & \times & X \times \mathbb{P}^1 & \to & X \times \mathbb{P}^1 \\ (t & , & (x, [x_0, x_1])) & & (x, [t^2 x_0, x_1]) \end{array} \cdot$$

From now on, in order to simplify calculations, we shall often consider cohomology with complex coefficients instead of rational ones; nevertheless, we keep the same notation, since it will be clear which field are we working on.

Recall that $H_T^*(Y) \cong H^*(X) \otimes \frac{\mathbb{C}[\hbar, p]}{(p^2 - p\hbar)}$, and that, localizing at fixed point components, we get:

$$H^*(X) \otimes \frac{\mathbb{C}(\hbar)[p]}{(p^2 - p\hbar)} \cong (H^*(X) \otimes \mathbb{C}(\hbar)) \oplus (H^*(X) \otimes \mathbb{C}(\hbar));$$

moreover, if $x \in H_T^*(Y)$ and $\delta(x) = (t, \tau)$, the equivariant integral is

$$\int_Y x = \frac{\int_X t - \int_X \tau}{\hbar};$$

finally, we denote $e_0 := \frac{p}{\hbar}$, $e_\infty := \frac{\hbar - p}{\hbar}$.

Now, let us fix a basis $\{\beta_1, ..., \beta_k\}$ of $H_2(X)$ and let β_0 be the generator of $H_2(\mathbb{P}^1)$, i.e. the dual of the hyperplane class; let $\{d = (d_1, ..., d_k), d_0\}$ be coordinates in $H_2(X \times \mathbb{P}^1)$. Insert "counters" $\{q = (q_1, ..., q_k), q_0\}$ in the equivariant potential, writing it as

$$\mathcal{F}(x) = \sum_{n \geq 3} \frac{1}{n!} \sum_{(d, d_0)} q_0^{d_0} q^d I_{(d, d_0)}^T \underbrace{(x, ..., x)}_{n},$$

and develop $\mathcal{F} = \sum_{i=0}^\infty q_0^i \mathcal{F}^i$.

Our goal is to prove that *solutions for the equation $\nabla_\hbar s = 0$ are given by second order derivatives of $\mathcal{F}^1$.*

3.2.1. Finding solutions of $\nabla_\hbar s = 0$.

Let us fix some notation: $Y_{k,(d,d_0)}$ is the "virtual" moduli space $\overline{\mathcal{M}}_{0,k}(X \times \mathbb{P}^1, (d, d_0))$: by this we mean that

$$\int_{Y_{k,(d,d_0)}}^T \omega := \int_{\overline{\mathcal{M}}_{0,k}(\mathbb{P}^n \times \mathbb{P}^1, (d, d_0))}^T \omega \cup \mathcal{E}_{k,(d,d_0)}^T,$$

where $\mathcal{E}_{k,(d,d_0)}^T = Euler^T(W_{k,(d,d_0)})$, and

$$W_{k,(d,d_0)}([C, x_1, ..., x_k, (\psi_1, \psi_2)]) := H^0(C, \psi_1^*(\oplus_{i=1}^r \mathcal{O}(l_i))).$$

Let F be the non-equivariant potential for X; we need to prove a technical lemma:

Lemma 3.5.

$$\mathcal{F}^0(x) = \frac{F(t) - F(\tau)}{\hbar}, \tag{3.1}$$

$$\partial_{e_0}\mathcal{F}^1(x) = -\partial_{e_\infty}\mathcal{F}^1(x) = \frac{1}{\hbar}\partial_p\mathcal{F}^1(x) = \frac{1}{\hbar}\mathcal{F}^1(x). \tag{3.2}$$

Proof. Since $\mathcal{F}^0(x) = \sum_{k \geq 3} \frac{1}{k!} \sum_d q^d I^T_{(d,0)} \left(\underbrace{x, ..., x}_{k} \right)$, we have to compute

$$
I^T_{(d,0)} \left(\underbrace{x, ..., x}_{k} \right) = \int_{Y_{k,(d,d_0)}}^T \rho_1^*(x) \cup ... \cup \rho_k^*(x) =
$$

$$
= \int_{\overline{\mathcal{M}}_{0,k}(\mathbb{P}^n \times \mathbb{P}^1, (d,0))}^T \rho_1^*(x) \cup ... \cup \rho_k^*(x) \cup \mathcal{E}^T_{k,(d,0)}.
$$

By definition, for every $[C, x_1, ..., x_k, (\psi_1, \psi_2)] \in \overline{\mathcal{M}}_{0,k}(\mathbb{P}^n \times \mathbb{P}^1, (d,0))$, the map ψ_2 is constant, and therefore $\overline{\mathcal{M}}_{0,k}(\mathbb{P}^n \times \mathbb{P}^1, (d,0))$ can be identified in a T equivariant way to $\overline{\mathcal{M}}_{0,k}(\mathbb{P}^n, d) \times \mathbb{P}^1$; the connected components of fixed points are $\overline{\mathcal{M}}_{0,k}(\mathbb{P}^n, d) \times \{0\}$ and $\overline{\mathcal{M}}_{0,k}(\mathbb{P}^n, d) \times \{\infty\}$.

Thus we can compute the GW invariants $I^T_{(d,0)}(x^k)$ by the localization theorem. By definition, $W_{k,(d,0)}$, restricted to the two copies of $\overline{\mathcal{M}}_{0,k}(\mathbb{P}^n, d)$, is $W_{k,d}$ itself. If we decompose $x = t + \tau$, and observe that $\rho_i^*(t)$ restricts to 0 on $\overline{\mathcal{M}}_{0,k}(\mathbb{P}^n, d) \times \{\infty\}$ and $\rho_i^*(\tau)$ restricts to 0 on $\overline{\mathcal{M}}_{0,k}(\mathbb{P}^n, d) \times \{0\}$, we get

$$
I^T_{(d,0)} \left(\underbrace{x, ..., x}_{k} \right) =
$$

$$
= \frac{\int_{\overline{\mathcal{M}}_{0,k}(\mathbb{P}^n, d)} \rho_1^*(t) \cup ... \cup \rho_k^*(t) \cup \mathcal{E}_{k,d} - \int_{\overline{\mathcal{M}}_{0,k}(\mathbb{P}^n, d)} \rho_1^*(\tau) \cup ... \cup \rho_k^*(\tau) \cup \mathcal{E}_{k,d}}{\hbar} =
$$

$$
= \frac{I_d(t^k) - I_d(\tau^k)}{\hbar}.
$$

From this the first formula follows .

Now write down $\mathcal{F}^1(x) = \sum_{k \geq 3} \frac{1}{k!} \sum_d q^d I^T_{(d,1)} \left(\underbrace{x, ..., x}_{k} \right)$, and observe that

$$
\partial_1 \mathcal{F}^1(x) = \sum_{k \geq 2} \frac{1}{k!} \sum_d q^d I^T_{(d,1)} \left(\underbrace{x, ..., x}_{k}, 1 \right) = 0 \text{ by properties of GW invariants,}
$$

since $(d, 1) \neq 0$.

We can see that

$$
\frac{1}{\hbar}\partial_p \mathcal{F}^1(x) = \frac{1}{\hbar}\sum_k \frac{1}{k!}\sum_d q^d I^T_{(d,1)}\left(\underbrace{x,...,x}_{k},p\right) =
$$

$$
= \frac{1}{\hbar}\sum_k \frac{1}{k!}\sum_d q^d \int p \cdot I^T_{(d,1)}\left(\underbrace{x,...,x}_{k}\right) =
$$

$$
= \frac{1}{\hbar}\sum_k \frac{1}{k!}\sum_d q^d I^T_{(d,1)}\left(\underbrace{x,...,x}_{k}\right) = \frac{1}{\hbar}\mathcal{F}^1(x).
$$

Furthermore, since $e_0 = \frac{p}{\hbar}$, we have

$$
\frac{1}{\hbar}\partial_p \mathcal{F}^1(x) = \frac{1}{\hbar}\sum_k \frac{1}{k!}\sum_d q^d I^T_{(d,1)}\left(\underbrace{x,...,x}_{k},p\right) =
$$

$$
= \sum_k \frac{1}{k!}\sum_d q^d \frac{1}{\hbar}I^T_{(d,1)}\left(\underbrace{x,...,x}_{k},p\right) =
$$

$$
= \sum_k \frac{1}{k!}\sum_d q^d I^T_{(d,1)}\left(\underbrace{x,...,x}_{k},\frac{p}{\hbar}\right) = \partial_{e_0}\mathcal{F}^1(x);
$$

By definition $e_0 + e_\infty = 1$, and therefore by a similar computation we get: $0 = \partial_1 \mathcal{F}^1(x) = \partial_{e_0}\mathcal{F}^1(x) + \partial_{e_\infty}\mathcal{F}^1(x) \Rightarrow \partial_{e_0}\mathcal{F}^1(x) = -\partial_{e_\infty}\mathcal{F}^1(x)$.

□

Let us take an orthonormal basis $\{T_0,...,T_m\}$ of $V \subseteq H^*(X)$; by the localization isomorphism, an orthogonal basis for the equivariant pairing in $H^*_T(X)$ is

$$
\left\{\frac{p}{\hbar}T_0, ..., \frac{p}{\hbar}T_m, \frac{\hbar-p}{\hbar}T_0, ..., \frac{\hbar-p}{\hbar}T_m\right\};
$$

the coordinates with respect to the first $m+1$ classes, the so called classes of type 0, are $\{t_i\}_{i=0,...,m}$, the coordinates of classes of type ∞ are $\{\tau_i\}_{i=0,...,m}$; each vector in the basis has squared norm $\frac{1}{\hbar}$.

Therefore, if $x = t+\tau$, this means that $t = \sum_{i=0}^m t_i \left(\frac{p}{\hbar}T_i\right)$ and $\tau = \sum_{i=0}^m \tau_i \left(\frac{\hbar-p}{\hbar}T_i\right)$.

Proposition 3.6. *The vector field* $\sum_{k=0}^m \frac{\partial^2 \mathcal{F}^1(x)}{\partial t_k \partial \tau_u}T_k$ *is a solution of the differential equation* $\nabla_\hbar s(t,\hbar) = 0$, *for every u.*

Proof. Let us write down the asociativity identity:

$$\sum_k \mathcal{F}_{xyk}\mathcal{F}_{kzu} = \sum_k \mathcal{F}_{xzk}\mathcal{F}_{kyu},$$

and take the coefficient of q_0:

$$\sum_k \mathcal{F}^0_{xyk}\mathcal{F}^1_{kzu} + \mathcal{F}^1_{xyk}\mathcal{F}^0_{kzu} = \sum_k \mathcal{F}^0_{xzk}\mathcal{F}^1_{kyu} + \mathcal{F}^1_{xzk}\mathcal{F}^0_{kyu}.$$

Next specialize $y = e_0 = \frac{p}{\hbar}T_0, x$ and z of type 0, u of type ∞, and notice from formula 3.1 that every time we differentiate $\mathcal{F}^0$ once by a variable of type 0, and once by a variable of type ∞, we get 0, therefore $\mathcal{F}^0_{kzu} = \mathcal{F}^0_{kyu} = 0$. Also $\mathcal{F}^1_{kyu} = \mathcal{F}^1_{kue_0} = \frac{1}{\hbar}\mathcal{F}^1_{ku}$ by 3.2, and

$$\mathcal{F}^0_{xyk} = \begin{cases} 0 \text{ if } k \text{ is of type } \infty \\ \frac{F_{xk1}(t)}{\hbar} = \frac{\langle x, t_k \rangle_X}{\hbar} \text{ if } k \text{ is of type } 0 \end{cases},$$

hence, on the left hand side we get $\frac{1}{\hbar}\sum_k \mathcal{F}^1_{kzu} \langle x, t_k \rangle_X = \frac{1}{\hbar}\mathcal{F}^1_{xzu}$
Hence substituting we obtain:

$$\frac{1}{\hbar}\mathcal{F}^1_{xzu} = \frac{1}{\hbar} \sum_k \frac{F_{xzk}}{\hbar}\mathcal{F}^1_{ku} = \frac{1}{\hbar^2} \sum_k F_{xzk}\mathcal{F}^1_{ku},$$

for every x, z of type 0, u of type ∞.
On the other side, $\nabla_\hbar \left(\sum_k \mathcal{F}^1_{ku}T_k \right) = 0$ if and only if

$$\hbar\frac{\partial \mathcal{F}^1_{xu}}{\partial tz}(t, \hbar) = \sum_{k=0}^m \mathcal{F}^1_{ku}(t, \hbar)\, F_{kzx}(t).$$

□

3.2.2. Separating variables.
We want to transform $\mathcal{F}^1_{ku}$ by mean of the localization formula in order to "separate" the dependence on the classes t and τ.

We have to understand, in the "virtual" space $Y_{k,(d,1)}$, the nature of the S^1-fixed points, that is, we have to understand the S^1 fixed points in $\overline{\mathcal{M}}_{0,k}(\mathbb{P}^n \times \mathbb{P}^1, (d, 1))$, and the behaviour of the Euler class.
From now on, assume $X = \mathbb{P}^n$, and then modify GW invariants with suitable Euler classes.
Remind that the fixed points have the following description (see figure 1):

$$Y_{k,(d,1)} = \left\{ \begin{array}{c} [C, x_1, ..., x_k, (\psi_1, \psi_2)] : \\ [C, x_1, ..., x_k, (\psi_1, \psi_2)] = [C, x_1, ..., x_k, (\psi_1, t\psi_2)] \,\forall t \in S^1 \end{array} \right\}.$$

This implies immediately that all the special points of C are mapped by ψ_2 on the fixed points of $\mathbb{P}^1$. In particular we have k_0 marked points mapping to 0 and k_∞ marked points mapping to ∞, $k = k_0 + k_\infty$.

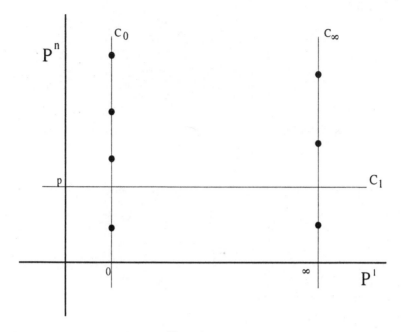

FIGURE 1.

Furthermore, by definition ψ_2 has degree 1, hence in general there is one special irreducible component C_1 of C which maps isomorphically to $\mathbb{P}^1$, while the image under ψ_2 of each one of the other components is a fixed point in $\mathbb{P}^1$. Thus the curve C is the union of three irreducible components C_1, C_0 and C_∞, such that:

$$
\begin{aligned}
\psi_{2|C_1} \quad &: \quad C_1 \xrightarrow{\sim} \mathbb{P}^1 \\
\psi_2(C_0) \quad &= \quad 0 \\
\psi_2(C_\infty) \quad &= \quad \infty.
\end{aligned}
$$

We claim that $\psi_{1|C_1}$ is constant: in fact, if $p \in C_1$ is generic and $\psi(p) = (x, a)$, then we obtain (x, ta) by acting with the group. For every $t \in S^1$, there must be a lifting of this action to C_1 which maps p to some other generic point of C_1 in $\psi_1^{-1}(x)$, hence $\psi_1^{-1}(x)$ contains an entire S^1-orbit, and therefore the whole C_1. It follows that $d = \deg \psi_1 = \deg \psi_{1|C_1} + \deg \psi_{1|C_0} + \deg \psi_{1|C_\infty} = 0 + d_0 + d_\infty$.
Finally remark that the marked points are necessarily on the two components (possibly degenerating to a point) C_0 and C_∞.

Definition 3.7.

$$M_{d_0,d_\infty,k_0,k_\infty} := \left\{ \begin{array}{c} [C,x_1,...,x_k,(\psi_1,\psi_2)] \in \left(Y_{k,(d,1)}\right)^T : \\ \psi_{2|C_1} : C_1 \longrightarrow \mathbb{P}^1, \psi_2\left(C_0\right) = 0, \psi_2\left(C_\infty\right) = \infty, \\ \deg \psi_{1|C_0} = d_0, \deg \psi_{1|C_\infty} = d_\infty, \\ \#\left\{\{x_1,...,x_k\} \cap C_0\right\} = k_0, \#\left\{\{x_1,...,x_k\} \cap C_\infty\right\} = k_\infty. \end{array} \right\}$$

We just proved:

Proposition 3.8.

$$\left(Y_{k,(d,1)}\right)^T = \cup_{\substack{d_0+d_\infty=d \\ k_0+k_\infty=k}} M_{d_0,d_\infty,k_0,k_\infty}.$$

In order to apply the localization formula we have now to compute:

- the Euler classes of the normal bundle of the components in the $M_{d_0,d_\infty,k_0,k_\infty}$'s,
- the restriction of the vector bundle $W_{k,(d,1)}$.

We denote by x_0 (resp. x_∞) the point $C_0 \cap C_1$ (resp. $C_\infty \cap C_1$); we shall say that C_0 (resp. C_∞) is degenerate if $\psi(x_0)$ (resp. $\psi(x_\infty)$) is smooth in the image of the curve; moreover, in this case $d_0 = 0$ (resp. $d_\infty = 0$).

We have to distinguish four cases:

1. Both C_0 and C_∞ are degenerate : this implies that $d = 0$, $k \leq 2$.
2. Only C_0 is degenerate; we have two cases according if $k_0 = 0$, $k_0 = 1$.
   - $k_0 = 0$; the variety $M_{0,d,0,k}$ is isomorphic to $X_{k+1,d}$:

$$\begin{array}{ccc} M_{0,d,0,k} & \to & X_{k+1,d} \\ [C,x_1,...,x_k,(\psi_1,\psi_2)] & & [C_\infty,x_\infty,x_1,...,x_k,\psi_1] \end{array}.$$

   - $k_0 = 1$; hence $x_0 = x_i$ for some $i \in \{1,...,k\}$, and the variety $M_{0,d,1,k-1}$ is a union of components isomorphic to $X_{k,d}$:

$$\begin{array}{ccc} M_{0,d,1,k-1}^i & \to & X_{k,d} \\ [C,x_1,...,x_k,(\psi_1,\psi_2)] & & [C_\infty,x_\infty,x_1,...,x_{i-1},x_{i+1},...,x_k,\psi_1] \end{array}.$$

3. Similarly for C_∞ degenerate:
   - $M_{d,0,k,0} \cong X_{k+1,d}$
   - $M_{d,0,k-1,1}^i \cong X_{k,d}$.
4. C_0, C_∞ are both non degenerate. For every partition of $\{1,...,k\}$ in two subsets A, B of cardinality k_0, k_∞, we get a component $M_{d_0,d_\infty,k_0,k_\infty}^{A,B}$, and the following map is an isomorphism:

$$\begin{array}{ccc} M_{d_0,d_\infty,k_0,k_\infty}^{A,B} & \to & X_{k_0+1,d_0} \times_X X_{k_\infty+1,d_\infty} \\ [C,x_1,...,x_k,(\psi_1,\psi_2)] & & \left[\left[C_0,x_0,x_{a_1},...,x_{a_{k_0}},\psi_1\right],\left[C_\infty,x_\infty,x_{b_1},...,x_{b_{k_\infty}},\psi_1\right]\right] \end{array},$$

where $X_{k_0+1,d_0} \overset{\rho_0}{\to} X$ and $X_{k_\infty+1,d_\infty} \overset{\rho_\infty}{\to} X$ are the evaluation maps on x_0 and x_∞.

Recall that, for every Z convex, on $Z_{k,d}$ we have line bundles F_i , $i = 1, ..., k$; by definition $F_i\,([C, x_1, ..., x_k, (\psi_1, \psi_2)]) = T_{x_i}C$, the tangent space to the curve C at the point x_i. In case of a group action on Z, these line bundle are canonically linearized by the pull back of the action on the tangent space at the point $\psi\,(x_i)$ to the image $\psi\,(C)$.

We shall denote by c_i the equivariant Chern class of F_i. All the same, in the case where C_0 and C_∞ are non degenerate, we get $F_0\,([C, x_1, ..., x_k, (\psi_1, \psi_2)]) := T_{x_0}C_0$, and $F_\infty\,([C, x_1, ..., x_k, (\psi_1, \psi_2)]) := T_{x_\infty}C_\infty$, with equivariant Chern classes c_0 and c_∞.

We want to compute the functions $\frac{\partial^2 \mathcal{F}^1(x)}{\partial t_k \partial \tau_u}$ by localization, thus we are only interested in mixed derivatives with a variable of type 0 and a variable of type ∞, which vanish in the cases where we have $k_0 = 0$ or $k_\infty = 0$; in fact, in these cases, the classes we should integrate to obtain GW invariants depend only on variables of type ∞ (or of type 0).

Remark 3.9. *Deformations of stable maps.*

We need to describe normal bundles to closed subvarieties in the moduli space of stable maps, hence we need a description of the tangent space to the moduli space $\overline{\mathcal{M}}_{0,k}\,(X, \beta)$ in terms of infinitesimal deformation of stable maps. For details and proofs see [FP] . Let $[C, x_1, ..., x_k, (\mu)]$ be a stable map, and let $Def\,(\mu)$ be the space of equivalence classes of first order infinitesimal deformations, i.e. the tangent space to $\overline{\mathcal{M}}_{0,k}\,(X, \beta)$ at $[C, x_1, ..., x_k, (\mu)]$. If $\{q_1, ...q_d\}$ are the nodes on the curve C, and C_i', C_i'' are the two branches of the curve meeting in q_i, we can decompose $Def\,(\mu) = Def_G\,(\mu) \oplus (\oplus_{i=1}^d T_{q_i}C_i' \otimes T_{q_i}C_i'')$; by $Def_G\,(\mu)$ we mean *infinitesimal deformations of the map which preserve the combinatorial type* (i.e. the dual graph) of the curve; there is a surjective map $Def_G\,(\mu) \to Def_G\,(C) \to 0$ on the space of infinitesimal deformations of the curve which preserve the combinatorial type, and we obtain the exact sequence

$$0 \to Def_C\,(\mu) \to Def_G\,(\mu) \to Def_G\,(C) \to 0;$$

by $Def_C\,(\mu)$ we denote the space of *infinitesimal deformations of the map whose restriction to the curve is trivial.* This space plays a role in another sequence:

$$0 \to H^0\left(C, TC\left(-\sum_{i=1}^d q_i\right)\right) \to Def_R\,(\mu) \to Def_C\,(\mu) \to 0;$$

here, $Def_R\,(\mu)$ is the space of *infinitesimal deformations of the map which keep the curve rigidly fixed*, and the kernel consists of reparametrization of the curve fixing the nodes. Looking at $Hom\,(C, X)$ as an open subset of the Hilbert scheme of graphs in $X \times C$, since $T_{[\mu]} \, Hom\,(C,X) \cong H^0\,(C, \mu^*\,(TX))$, we build the last exact sequence in order to obtain information on $Def_R\,(\mu)$:

$$0 \to \oplus_{j=1}^k Def_i\,(\mu) \to Def_R\,(\mu) \to H^0\,(C, \mu^*\,(TX)) \to 0;$$

$Def_j\,(\mu) \cong T_{x_j} C$ is the infinitesimal deformation which just moves the j−th marking.

Proposition 3.10. *Let D be a connected component of fixed points in $Y_{k,(d,1)}$. Then*

$$Euler\left(\mathcal{N}_{D/Y_{k,(d,1)}}\right) =$$

$$= \begin{cases} -\hbar^2 & \text{if } D \subset M_{0,0,1,1}\,(1) \\ -\hbar^2\,(-\hbar + c_\infty) & \text{if } D \subset M_{0,d,1,k-1}\,(2) \\ -\hbar^2\,(\hbar + c_0) & \text{if } D \subset M_{d,0,k-1,1}\,(3) \\ -\hbar^2\,(\hbar + c_0)\,(-\hbar + c_\infty) & \text{if } D \subset M_{d_0,d_\infty,k_0,k_\infty},\ d_0, d_\infty \neq 0\ (4) \end{cases}$$

Proof.
Case 1) $Y_{2,(0,0)} \cong X \times \mathbb{P}^1 \times \mathbb{P}^1$, T -equivariantly. The isomorphism is

$$[C, x_0, x_\infty, (\psi_1, \psi_2)] \to (\psi_1\,(C), \psi_2\,(x_0), \psi_2\,(x_\infty)),$$

and the two connected components of fixed points we need to consider, say D and D' corresponds to $X \times \{0\} \times \{\infty\}$ and $X \times \{\infty\} \times \{0\}$. The linearized normal bundle at some point $\{x\} \times \{0\} \times \{\infty\}$ is identified with the two dimensional vector space $\{x\} \times \mathbb{P}^1 \setminus \{\infty\} \times \mathbb{P}^1 \setminus \{0\}$, and S^1 acts with characters $(-\hbar, \hbar)$, hence the equivariant Euler class is $-\hbar^2$; similarly, S^1 acts with characters $(\hbar, -\hbar)$ on the normal bundle at some point $\{x\} \times \{\infty\} \times \{0\}$, and the Euler class is still $-\hbar^2$.

Case 2) Consider the following subvarieties in $Y_{k,(d,1)}$:

$$E_i = \left\{ \begin{array}{c} [C, x_1, ..., x_k, (\psi_1, \psi_2)] : C = C_1 \cup C_2, \\ \deg \psi_{2|C_1} = 1, \deg \psi_{2|C_2} = 0, \\ \{\{x_1, ..., x_k\} \cap C_1\} = \{x_i\} \end{array} \right\}.$$

The preimage of the divisor $D_{\{i\}\{1,...,i-1,i+1,...,k\}}$ under the map $Y_{k,(d,1)} \to \mathcal{M}_{0,k}$, $[C, x_1, ..., x_k, (\psi_1, \psi_2)] \to [C, x_1, ..., x_k]$, consists of two components, according to $\deg \psi_{2|C_1} = 1$ or $\deg \psi_{2|C_1} = 0$; the first one is exactly E_i, hence it is a divisor in $Y_{k,(d,1)}$.
This divisor maps equivariantly to $\mathbb{P}^1 \times \mathbb{P}^1$:

$$\begin{array}{ccc} E_i & \to & \mathbb{P}^1 \times \mathbb{P}^1 \\ [C, x_1, ..., x_k, (\psi_1, \psi_2)] & & (\psi_2\,(x_i), \psi_2\,(x_\infty)) \end{array},$$

and the connected components of fixed points $M_{0,d,1,k-1}^i$ are the preimages of the point $(0, \infty)$. By the same reasoning of the previous case, $\mathcal{N}_{M_{0,d,1,k-1}^i/E_i} = -\hbar^2$.
In order to obtain $\mathcal{N}_{M_{0,d,1,k-1}^i/Y_{k,(d,1)}}$, we have to multiply by $Euler\left(\mathcal{N}_{E_i/Y_{k,(d,1)}}\right)$.

By the remark 3.9, the fiber of the bundle $\mathcal{N}_{E_i/Y_{k,(d,1)}}$ is the tensor product of the two tangent spaces to the two irreducible components of the curve in the join point, since the combinatorial type of the curve is fixed in E_i.

In our case, the tangent space to C_1 at the point x_i is acted on by S^1 with character $-\hbar$, the line bundle on E_i is trivial with equivariant Chern class ; furthermore the tangent space to C_∞ at x_∞ has trivial action, hence the equivariant Chern class of the line bundle is c_∞. Thus $Euler\left(\mathcal{N}_{E_i/Y_{k,(d,1)}}\right) = -\hbar + c_\infty$, and

$$Euler\left(\mathcal{N}_{D/Y_{k,(d,1)}}\right) = -\hbar^2\left(-\hbar + c_\infty\right).$$

Case 3) The computation is exactly the same as in case 2, except that the action on the tangent space to C_1 at the point x_i has character $\hbar$, and the equivariant Chern class of the line bundle of the tangent space to C_0 at x_0 is c_0, hence

$$Euler\left(\mathcal{N}_{D/Y_{k,(d,1)}}\right) = -\hbar^2\left(\hbar + c_0\right).$$

Case 4) Again, consider the map $\pi : Y_{k,(d,1)} \to \overline{\mathcal{M}}_{0,k}$; for every partition of $\{1, ..., k\}$ in two subsets A, B of cardinality k_0, k_∞, $\pi^{-1}(D_{A,B})$ is a divisor, and contains as codimension 1 subvariety

$$E_{A,B} = \left\{ \begin{array}{c} [C, x_1, ..., x_k, (\psi_1, \psi_2)] \in \pi^{-1}(D_{A,B}) : C = C_1 \cup C_2 \cup C_3, \\ \deg \psi_{2|C_1} = 1, \deg \psi_{2|C_2} = \deg \psi_{2|C_3} = 0 \end{array} \right\},$$

which can be seen as the complete intersection of divisors

$$\left\{ \begin{array}{c} [C, x_1, ..., x_k, (\psi_1, \psi_2)] \in \pi^{-1}(D_{A,B}) \\ \deg \psi_{2|C_1} = 1, \deg \psi_{2|C_2} = 0 \end{array} \right\} \cap \left\{ \begin{array}{c} [C, x_1, ..., x_k, (\psi_1, \psi_2)] \in \pi^{-1}(D_{A,B}) \\ \deg \psi_{2|C_1} = 0, \deg \psi_{2|C_2} = 1 \end{array} \right\}.$$

Hence, by the same computations, $Euler\left(\mathcal{N}_{E_{A,B}/Y_{k,(d,1)}}\right) = (-\hbar + c_\infty)(\hbar + c_0)$. Analogously, we get an equivariant map

$$\begin{array}{ccc} E_{A,B} & \to & \mathbb{P}^1 \times \mathbb{P}^1 \\ [C, x_1, ..., x_k, (\psi_1, \psi_2)] & & (\psi_2(x_0), \psi_2(x_\infty)) \end{array},$$

and our connected component of fixed points is the preimage of $(0, \infty)$, hence has normal bundle in $E_{A,B}$ with equivariant Euler class $-\hbar^2$. Thus $Euler\left(\mathcal{N}_{D/Y_{k,(d,1)}}\right) = -\hbar^2(\hbar + c_0)(-\hbar + c_\infty)$.

$\square$

Recall that by definition we have

$$\frac{\partial^2 \mathcal{F}^1(x)}{\partial t_i \partial \tau_j} = \sum_{k \geq 1} \frac{1}{k!} \sum_d q^d I_{(d,1)}^T\left(x^k, T_i\frac{p}{\hbar}, T_j\frac{\hbar - p}{\hbar}\right),$$

hence, in order to separate variables, we apply the integration formula to

$$I_{(d,1)}^T\left(x^k, T_i\frac{p}{\hbar}, T_j\frac{\hbar - p}{\hbar}\right).$$

First of all, let us fix a suitable basis for $H^*(X)$; let $v := Euler\left(\oplus_{i=1}^r \mathcal{O}(l_i)\right)$, and let $\{T_0, ..., T_s\}$ be a set of independent elements such that $\{[T_0], ..., [T_s]\}$ form an orthonormal basis in $\frac{H^*(X)}{Ann(v)}$, hence $\int_X T_i \cup T_j v = \delta_{ij}$. Let us denote $T_i v$ with T_i' and complete to orthonormal dual basis of $H^*(X)$, $\{T_0, ..., T_s, T_{s+1}, ..., T_m\}$ and $\{T_0', ..., T_s', T_{s+1}', ..., T_m'\}$, such that the class of the diagonal in X is

$$\Delta = \sum_{i=0}^m T_i \otimes T_i'.$$

We also define $W_{k+1,d}'$ as the kernel of the map

$$W_{k+1,d} \quad \rightarrow \quad \rho_1^*\left(\oplus_{i=1}^r \mathcal{O}(l_i)\right)$$
$$\sigma \quad \rightarrow \quad \sigma\left(\psi\left(x_{k+1}\right)\right),$$

therefore $\mathcal{E}_{k+1,d}' := Euler(W_{k+1,d}')$ satisfies $\mathcal{E}_{k+1,d} = \mathcal{E}_{k+1,d}' \rho_{k+1}^*(v)$.

Lemma 3.11. Let $D = M_{d_0,d_\infty,n_0,n_\infty}^{A,B}$, then

$$\int_D \frac{\rho_1^*(x)...\rho_k^*(x)\,\mathcal{E}_{k,(d,1)}}{Euler\left(\mathcal{N}_{D/Y_{k,(d,1)}}\right)} =$$

$$= \sum_{i=0}^m \int_{X_{k_0+1,d_0}} \rho_1^*(t)...\rho_{k_0}^*(t)\,\frac{\rho_0^*(T_i)\,\mathcal{E}_{k_0+1,d}}{\hbar(\hbar+c_0)}.$$

$$\cdot \int_{X_{k_\infty+1,d_\infty}} \rho_1^*(\tau)...\rho_{k_\infty}^*(\tau)\,\frac{\rho_\infty^*(T_i)\,\mathcal{E}_{k_\infty+1,d}}{\hbar(-\hbar+c_\infty)}.$$

Proof. We have $\binom{k}{k_0}$ such components in $M_{k_0,k_\infty,d_0,d_\infty}$, on varying the sets A and B, and each one gives the same contribution, thus we assume without loss of generality that $A = \{1, ..., k_0\}$; recall that we have the isomorphism:

$$M_{d_0,d_\infty,k_0,k_\infty}^{A,B} \quad \xrightarrow{\phi} \quad X_{k_0,d_0+1} \times_X X_{k_\infty,d_\infty+1}$$
$$[C, x_1, ..., x_k, (\psi_1, \psi_2)] \quad [[C_0, x_0, x_1, ..., x_{k_0}, \psi_1], [C_\infty, x_\infty, x_{k_0+1}, ..., x_k, \psi_1]]$$

As a first step, let us compute the restriction of $\mathcal{E}_{k,d}$ to this component. Let ν be the evaluation map on the meeting point, then we get the exact sequence:

$$0 \quad \rightarrow \quad W_{k_0+1,d_0}' \oplus W_{k_\infty+1,d_\infty}' \quad \rightarrow \quad W_{k,(d,1)} \quad \rightarrow \quad \nu^*\left(\oplus_i \mathcal{O}(l_i)\right) \quad \rightarrow \quad 0$$
$$[\psi_1^*(\sigma), \psi_1^*(\tau)] \quad \rightarrow \quad \sigma(x_0) = \tau(x_\infty)$$

Therefore $\mathcal{E}_{k,(d,1)} = \mathcal{E}_{k_0+1,d_0}' \mathcal{E}_{k_\infty+1,d_\infty}' \nu^*(v)$.

Putting everything toghether, and denoting with $\mu = (\rho_0, \rho_\infty)$ the product of evaluation maps on the last marked points, $X_{k_0,d_0+1} \times X_{k_\infty,d_\infty+1} \to X \times X$, we get

$$\int_D \frac{\rho_1^*(x)...\rho_k^*(x)\,\mathcal{E}_{k,(d,1)}}{Euler\left(\mathcal{N}_{D/Y_{k,(d,1)}}\right)} =$$

$$= \int_D \frac{\rho_1^*(t)...\rho_{k_0}^*(t)\,\rho_{k_0+1}'(\tau)...\rho_k^*(\tau)\,\mathcal{E}_{k_0+1,d_0}'\mathcal{E}_{k_\infty+1,d_\infty}'\mu^*\,(v\,(1\otimes 1))\,\mu^*\,(\Delta)}{-\hbar^2\,(\hbar + c_0)\,(-\hbar + c_\infty)} =$$

$$= \int_D \frac{\rho_1^*(t)...\rho_{k_0}^*(t)\,\rho_{k_0+1}'(\tau)...\rho_k^*(\tau)\,\mathcal{E}_{k_0+1,d_0}'\mathcal{E}_{k_\infty+1,d_\infty}'\mu^*\,(v\sum_{i=0}^m T_i\otimes T_i')}{-\hbar^2\,(\hbar + c_0)\,(-\hbar + c_\infty)} =$$

$$= \int_D \frac{\rho_1^*(t)...\rho_{k_0}^*(t)\,\rho_{k_0+1}'(\tau)...\rho_k^*(\tau)\,\mathcal{E}_{k_0+1,d_0}'\mathcal{E}_{k_\infty+1,d_\infty}'\mu^*\,((v\otimes v)\sum_{i=0}^m T_i\otimes T_i)}{-\hbar^2\,(\hbar + c_0)\,(-\hbar + c_\infty)} =$$

$$= -\frac{1}{\hbar^2}\sum_i \int_{X_{k_0,d_0+1}} \frac{\rho_1^*(t)...\rho_{k_0}^*(t)\,\rho_0^*(T_i)\,\mathcal{E}_{k_0+1,d_0}}{(\hbar + c_0)}\cdot$$

$$\cdot \int_{X_{k_\infty,d_\infty+1}}^T \frac{\rho_1^*(\tau)...\rho_{k_\infty}^*(\tau)\,\rho_\infty^*(T_i)\,\mathcal{E}_{k_\infty+1,d_\infty}}{(-\hbar + c_\infty)}$$

□

With similar arguments we compute the contribution of components of type 2 and 3; for example, if $D = M_{d,0,k-1,1}^j$, then

$$\int_D \frac{\rho_1^*(x)...\rho_k^*(x)\,\mathcal{E}_{k,(d,1)}}{Euler\left(\mathcal{N}_{D/Y_{k,(d,1)}}\right)} = -\frac{1}{\hbar^2}\sum_i \int_{X_{k,d}} \frac{\rho_1^*(t)...\rho_{i-1}^*(t)\,\rho_{i+1}^*(t)...\rho_k^*(t)\,\rho_0^*(T_i)\,\mathcal{E}_{k,d}}{(\hbar + c_0)}.$$

By definition, we set $\int_{X_{2,0}} \frac{\rho_1^*(T_i)\rho_2^*(T_j)}{\hbar + c} := \langle T_i, T_j\rangle$.

We now set

Definition 3.12.

$$\psi_{ij}(t,\hbar) = \sum_{k=0}^\infty \frac{1}{k!}\sum_{d=0}^\infty q^d \int_{X_{k+2,d}} \frac{\rho_1^*(t)...\rho_k^*(t)\,\rho_{k+1}^*(T_i)\,\rho_{k+2}^*(T_j)\,\mathcal{E}_{k+2,d}}{(\hbar + c)}.$$

Now, lemma 3.11 easily implies:

Theorem 3.13.

$$-\hbar^2\frac{\partial^2\mathcal{F}^1(x)}{\partial t_i\partial\tau_j} = \sum_k \psi_{ik}(t,\hbar)\,\psi_{kj}(\tau,-\hbar)$$

Proof. Take coordinates $\{t_0,...,t_m,\tau_0,...,\tau_m\}$ in $H_T^*(X)$ relative to the basis $\left\{T_0\frac{p}{\hbar},...,T_m\frac{p}{\hbar}, T_0\frac{\hbar-p}{\hbar},...,T_m\frac{\hbar-p}{\hbar}\right\}$.

Recall that

$$\frac{\partial^2\mathcal{F}^1(x)}{\partial t_i\partial\tau_j} = \sum_{k\geq 1}\frac{1}{k!}\sum_d q^d I_{(d,1)}^T\left(x^k, T_i\frac{p}{\hbar}, T_j\frac{\hbar-p}{\hbar}\right).$$

Summing up contributions of components of type 1, 2, 3, 4,

$$I_{(d,1)}^T\left(x^k, T_i\frac{p}{\hbar}, T_j\frac{\hbar-p}{\hbar}\right) =$$

$$= -\frac{1}{\hbar^2}\langle T_i, T_j\rangle - \frac{1}{\hbar^2}k\sum_l\int_{X_{k+2,d}}\frac{\rho_1^*(t)...\rho_k^*(t)\,\rho_{k+1}^*(T_i)\,\rho_{k+2}^*(T_l)\,\mathcal{E}_{k,d}}{(\hbar+c_0)}\langle T_l,T_j\rangle +$$

$$-\frac{1}{\hbar^2}k\sum_l\int_{X_{k+2,d}}\frac{\rho_1^*(\tau)...\rho_k^*(\tau)\,\rho_{k+1}^*(T_j)\,\rho_{k+2}^*(T_l)\,\mathcal{E}_{k,d}}{(-\hbar+c_\infty)}\langle T_l,T_i\rangle +$$

$$-\frac{1}{\hbar^2}\sum_{\substack{d_0,d_\infty>0\\k_0,k_\infty>0}}\binom{k}{k_0}\sum_l\int_{X_{k_0+2,d}}\frac{\rho_1^*(t)...\rho_{k_0}^*(t)\,\rho_{k_0+1}^*(T_i)\,\rho_{k_0+2}^*(T_l)\,\mathcal{E}_{k_0+2,d_0}}{(\hbar+c_0)}\cdot$$

$$\cdot\int_{X_{k_\infty+2,d}}\frac{\rho_1^*(\tau)...\rho_{k_\infty}^*(\tau)\,\rho_{k_\infty+1}^*(T_l)\,\rho_{k_\infty+2}^*(T_j)\,\mathcal{E}_{k_\infty+2,d_\infty}}{(-\hbar+c_\infty)}.$$

Substituting in the formula for $\frac{\partial^2 \mathcal{F}^1(x)}{\partial t_i\partial\tau_j}$, and recalling that $\{T_0,...,T_m\}$ is an orthonormal basis, the result follows.

$\square$

Theorem 3.14. *The matrix $\Psi(t,\hbar)$ is a fundamental matrix of solutions for the differential equation*

$$\nabla_\hbar s(x,t) = 0.$$

Proof. $\left(\frac{\partial^2\mathcal{F}^1(x)}{\partial t_i\partial\tau_j}\right) = -\frac{1}{\hbar^2}\Psi(t,\hbar)\,\Psi(\tau,-\hbar)$ is a fundamental matrix of solutions. Since the part of degree 0 in q of $\Psi(\tau,-\hbar)$ is $-\frac{1}{\hbar^2}\langle T_i,T_j\rangle$, this matrix is invertible, and we can write

$$\Psi(t,\hbar) = -\hbar^2\left(\frac{\partial^2\mathcal{F}^1(x)}{\partial t_i\partial\tau_j}\right)\Psi^{-1}(\tau,-\hbar).$$

Since $\Psi^{-1}(\tau,-\hbar)$ does not depend on t, and we are differentiating with respect to the variables $\{t_i\}$, we have that $\Psi(t,\hbar)$ is another fundamental matrix of solutions.

$\square$

In general, given any basis $\{v_0,...,v_n\}$ for $H^*(X)$, with g_{ij} the corresponding intersection matrix, we can define

$$\psi_{ij}(t,\hbar) = \sum_{k=0}^{\infty}\frac{1}{k!}\sum_{d=0}^{\infty}q^d\int_{X_{k+2,d}}\frac{\rho_1^*(t)...\rho_k^*(t)\,\rho_{k+1}^*(v_i)\,\rho_{k+2}^*(v_j)\,\mathcal{E}_{k+2,d}}{(\hbar+c)},$$

and then prove in a similar way that

$$-\hbar^2\frac{\partial^2\mathcal{F}^1(x)}{\partial t_i\partial\tau_j} = \sum_{k,l}\psi_{ik}(t,\hbar)\,g^{kl}\psi_{lj}(\tau,-\hbar).$$

This implies that theorem 3.14 holds for any choice of the basis.

A quicker way to prove this is to observe that for any $\omega \in H^* (X)$, the element

$$\psi (\omega, t, \hbar) \quad : \quad =$$

$$= \sum_{i=0}^{m} \left(\sum_{k=0}^{\infty} \frac{1}{k!} \sum_{d=0}^{\infty} q^d \int_{X_{k+2,d}} \frac{\rho_1^* (t) ... \rho_k^* (t) \, \rho_{k+1}^* (T_i) \, \rho_{k+2}^* (\omega) \, \mathcal{E}_{k+2,d}}{(\hbar + c)} \right) T_i$$

is a solution of the differential equation $\nabla_\hbar s(x, t) = 0$, because of the linearity of $\psi (\omega, t, \hbar)$ in the variable ω. From now on we will use a homogeneous basis with respect to the grading, and denote the unity by T_0.

3.2.3. Restriction to $SQH^* (X)$.

We now pass to study how our solution $\Psi (t, \hbar)$ restricts for $t = t_1 T_1$, i.e. what happens if we consider our connections restricted to the tangent bundle to $H^2 (X)$. We recall that by our assumptions $H^2 (X)$ (or $H^2 (X) \cap V$) is one-dimensional. The solutions of $\nabla_\hbar = 0$ will be considered as elements in the small quantum cohomology ring of X.
Our aim is to prove:

Theorem 3.15. *Let* $s_{ij} (t, \hbar) := \psi_{ij} (t, \hbar)_{|t \in H^2(X)}$. *Then, for every* i, j, *we have*

$$s_{ij} (t, \hbar) = \sum_{d \geq 0} q^d \exp (dt_1) \int_{X_{2,d}} \exp \left(\frac{\rho_1^* (t)}{\hbar} \right) \frac{\rho_1^* (T_i) \, \rho_2^* (T_j)}{\hbar + c} \mathcal{E}_{2,d}.$$

The proof of this theorem requires some simple facts. Consider the map

$$\pi \quad : \quad X_{k,d} \to X_{k-1,d}$$
$$[C, x_1, ..., x_k, \psi] \quad \to \quad [C, x_1, ..., x_{k-1}, \psi]$$

which forgets the last point, and the map $\sigma_1 : X_{k-1,d} \to X_{k,d}$, the section given by the first marked point. Let $D := \sigma_1 (X_{k-1,d})$, and $L := \mathcal{O} (D)$. Set F_1^k equal to the line bundle on $X_{k,d}$ whose fibers are the tangent lines at the first marked points.

Lemma 3.16.   1. $\sigma_1^* (F_1^k)$ *is trivial.*
    2. $\pi^* (F_1^{k-1}) = F_1^k \otimes L.$

Proof. 1) By definition of the section σ_1, the divisor D consists of those stable maps $[C, x_1, ..., x_k, \psi]$ such that C splits in two irreducible components C_1 and C_2, such that C_1 contains exactly the two marked points x_1 and x_k, and ψ restricted to C_1 is constant. This immediately implies our claim since the line bundle $\sigma_1^* (F_1^k)$ is the pull back of a bundle on $\overline{\mathcal{M}}_{0,3} = \{pt\}$.
2) It is clear that $\pi^* (F_1^{k-1})$ and F_1^k are canonically isomorphic outside D; thus $\pi^* (F_1^{k-1}) \otimes (F_1^k)^{-1} = L^a$, for some a, since it has support only on D. To compute the integer a, let us restrict to D. We have already used that the fiber of $L_{|D} = \mathcal{N}_{D/X_{k,d}}$ is $T_p C_1 \otimes T_p C_2$, where p is the join point.

In this case, the first factor is trivial, and the second equals $\pi^* \left(F_1^{k-1} \right)$, hence $L_{|D} = \pi^* \left(F_1^{k-1} \right)$.

On the other hand, by 1) we deduce that $\left(F_1^k \right)_{|D} = \sigma_1^* \left(F_1^k \right)$ is trivial. Substituting, we get $\left(\pi^* \left(F_1^{k-1} \right) \otimes \left(F_1^k \right)^{-1} \right)_{|D} = \pi^* \left(F_1^{k-1} \right)$, and $a = 1$.

$\square$

We set $c(k) := c_1 \left(F_1^k \right) \in H^2 \left(X_{k,d} \right)$; with abuse of notation, we will write only c if it is clear to which cohomology ring it belongs.

Lemma 3.17. *Let $\pi : X_{k,d} \to X_{k-1,d}$ be the map which forgets the last marked point, and let $a \in H^* \left(X_{k,d} \right)$. Then for any $r \geq 1$,*

$$\pi_* \left(ac^r \right) = \pi_* a \left(c^r \right) - \sigma_1^* \left(a \right) c^{r-1}.$$

Proof. From the previous lemma , if we set $x := \sigma_* (1) = c_1(L)$, we get:

$$\begin{cases} \sigma_1^* \left(c^r \right) = 0 \\ \pi^* \left(c(k-1) \right) = c(k) + x \end{cases} .$$

For $r = 1$:

$$\begin{aligned} \pi_* \left(ac(k) \right) & = \pi_* \left\{ a \left[\pi^* c(k-1) - x \right] \right\} = \pi_* \left[a\pi^* c(k-1) \right] - \pi_* (ax) = \\ & = \pi_* (a) c(k-1) - \pi_* \left(a\sigma_{1*} (1) \right) = \pi_* (a) c(k-1) - \pi_* \sigma_{1*} \left(1 \cdot \sigma_1^* (a) \right) = \\ & = \pi_* (a) c(k-1) - \sigma_1^* (a) . \end{aligned}$$

By induction, if it is true for $r - 1$,

$$\begin{aligned} \pi_* \left(ac^r \right) & = \pi_* \left(ac^{r-1} c \right) = \pi_* \left[ac^{r-1} \left(\pi^* c - x \right) \right] = \pi_* \left(ac^{r-1} \right) c - \pi_* \left[ac^{r-1} \sigma_{1*} (1) \right] = \\ & = \pi_* \left(ac^{r-1} \right) c - \pi_* \sigma_{1*} \left[1 \cdot \sigma_1^* \left(ac^{r-1} \right) \right] = \pi_* \left(ac^{r-1} \right) c - \sigma_1^* \left(ac^{r-1} \right) = \\ & = \pi_* a \left(c^r \right) - \sigma_1^* (a) c^{r-1} - \sigma_1^* \left(ac^{r-1} \right) = \pi_* a \left(c^r \right) - \sigma_1^* (a) c^{r-1}. \end{aligned}$$

$\square$

Proof of 3.15. Recall that

$$s_{ij}(t, \hbar) = \psi_{ij} (t, \hbar)_{|t \in H^2(X)} , \text{ and}$$

$$\psi_{ij} (t, \hbar) = \sum_{k=0}^{\infty} \frac{1}{k!} \sum_{d=0}^{\infty} q^d \int_{X_{k+2,d}} \frac{\rho_1^* (T_i) \rho_2^* (T_j) \rho_3^* (t) \dots \rho_{k+2}^* (t) \mathcal{E}_{k+2,d}}{(\hbar + c)}.$$

Let $a \in H^* (X)$. Developing in power series, we have

$$\frac{a}{\hbar + c} = \sum_{r=0}^{\infty} \frac{(-1)^r}{\hbar^{r+1}} ac^r,$$

and

$$\pi_* \left(\frac{a}{\hbar + c} \right) = \sum_{r=0}^{\infty} \frac{(-1)^r}{\hbar^{r+1}} \pi_* (ac^r) = \sum_{r=0}^{\infty} \frac{(-1)^r}{\hbar^{r+1}} \left[\pi_* (a) (c^r) - \sigma_1^* (a) c^{r-1} \right] =$$

$$= \frac{\pi_* (a)}{\hbar + c} - \frac{\sigma_* (a)}{\hbar (\hbar + c)}.$$

Moreover, if $y \in H^* (X_{k,d})$, and $y' = \pi_* (y)$, then

$$\pi_* \left(\frac{y \rho_k^* (t)}{\hbar + c} \right) = \frac{\pi_* (y \rho_k^* (t))}{\hbar + c} - \frac{\sigma_* (y \rho_k^* (t))}{\hbar (\hbar + c)} =$$

$$= \frac{y' \pi_* \rho_k^* (t)}{\hbar + c} - \frac{\sigma_* \rho_k^* (t)}{\hbar} \frac{y'}{\hbar + c} =$$

$$= \frac{y' \langle d, t \rangle}{\hbar + c} - \frac{\rho_1^* (t)}{\hbar} \frac{y'}{\hbar + c}.$$

Using that $\mathcal{E}_{k,d} = \pi^* (\mathcal{E}_{k-1,d})$, apply this last formula to obtain:

$$\int_{X_{k+2,d}} \frac{\rho_1^* (T_i) \rho_2^* (T_j) \rho_3^* (t) ... \rho_{k+2}^* (t) \mathcal{E}_{k+2,d}}{(\hbar + c)} =$$

$$= \int_{X_{k+1,d}} \frac{\pi_* (\rho_1^* (T_i) \rho_2^* (T_j) \rho_3^* (t) ... \rho_{k+2}^* (t)) \mathcal{E}_{k+1,d}}{(\hbar + c)} =$$

$$= \int_{X_{k+1,d}} \frac{\rho_1^* (T_i) \rho_2^* (T_j) \rho_3^* (t) .. \rho_{k+1}^* (t)}{\hbar + c} \left(dt_1 - \frac{\rho_1^* (t)}{\hbar} \right) \mathcal{E}_{k+1,d} =$$

$$= ... = \int_{X_{2,d}} \frac{\rho_1^* (T_i) \rho_2^* (T_j)}{\hbar + c} \left(dt_1 - \frac{\rho_1^* (t)}{\hbar} \right)^k \mathcal{E}_{2,d} =$$

$$= \int_{X_{2,d}} \frac{\rho_1^* (T_i) \rho_2^* (T_j)}{\hbar + c} \sum_{u=0}^{k} \binom{k}{u} (dt_1)^u \left(\frac{\rho_1^* (t)}{\hbar} \right)^{k-u} \mathcal{E}_{2,d}.$$

Hence

$$s_{ij} (t, \hbar) = \sum_{d=0}^{\infty} q^d \sum_{k=0}^{\infty} \frac{1}{k!} \sum_{u=0}^{k} \binom{k}{u} (dt_1)^u \int_{X_{2,d}} \frac{\rho_1^* (T_i) \rho_2^* (T_j)}{\hbar + c} \left(\frac{\rho_1^* (t)}{\hbar} \right)^{k-u} \mathcal{E}_{2,d} =$$

$$= \sum_{d \geq 0} q^d \exp (dt_1) \int_{X_{2,d}} \exp \left(\frac{\rho_1^* (t)}{\hbar} \right) \frac{\rho_1^* (T_i) \rho_2^* (T_j)}{\hbar + c} \mathcal{E}_{2,d}.$$

$\square$

A special role will be played by the collection of functions

$$s_\beta (t, \hbar) := s_{0\beta} (t, \hbar),$$

i.e. by the first components of the vector solutions of $\nabla_\hbar s = 0$.

These will be the functions to manipulate in order to get solutions for Picard Fuchs equation, and these give enough information about enumerative geometry in X.

Let $P(\hbar \frac{d}{dt_1}, \exp t_1, \hbar)$ denote a non-commutative polynomial differential operator in the variables $\hbar, \frac{d}{dt_1}, \exp t_1$, that is

$$P(\hbar \frac{d}{dt_1}, \exp t_1, \hbar) = \sum_r Q_r(\exp t_1, \hbar) \left(\hbar \frac{d}{dt_1} \right)^r,$$

Proposition 3.18. *Take* $P(\hbar \frac{d}{dt_1}, \exp t_1, \hbar)$ *as above. If*

$$Ps_\beta (t, \hbar) = 0$$

for every $\beta = 0, ..., m$, *then the following relation holds in* $SQH^* (X)$:

$$P(T_1, q, 0) = 0.$$

Proof. The functions s_β form the first row in the fundamental solution matrix $S = \{s_{ij}\}$. Let M_1 be the matrix of quantum multiplication by the class T_1. Obviously,

$$(M_1 \cdot S)_{ij} = \sum_k s_{kj} \phi_{1ki} = \hbar \frac{d}{dt_1} s_{ij};$$

on the other hand, as functions ϕ_{1ki} depend on t_1, we have

$$\left(\hbar \frac{d}{dt_1} \right)^r s_{ij} = T_1^r * s_{ij} + \hbar \cdot \tilde{Q},$$

where $\tilde{Q}$ is the multiplication by a matrix whose entries are functions of the variable T_1 and depend polynomially on $\hbar$.

Develop the polynomial P with respect to the last variable, i.e. $P = P_0 + \hbar P_1 + ... + \hbar^r P_r$, and

$$P = P_0 + \hbar P_1 + ... + \hbar^r P_r,$$

reducing modulo $\hbar$, we get:

$$(P_0(M_1, \exp t_1) \cdot S)_{0\beta} = (P(M_1, \exp t_1, 0) \cdot S)_{0\beta} = 0.$$

We know that S is a non-degenerate matrix, and therefore

$$(P(M_1, \exp t_1, 0))_{0\beta} = 0,$$

for every $\beta = 0, ..., m$.

Next goal is to prove that the first column of the matrix $P(M_1, \exp t_1, 0)$ vanishes. This would imply that this quantum operator, applied to the unit T_0, vanishes, and therefore $P(T_1, q, 0)$ in $SQH^* (X)$. Since we can choose an orthogonal basis, for every index i, writing down explicitly the expression of $(P(M_1, \exp t_1, 0)) T_0 * T_1$ and using the fact that T_0 is a unit for the quantum product, we obtain that the i-th element of the first column of our matrix is 0.□

Part 4. Solutions of Picard-Fuchs equation

From now on let X be a smooth projective complete intersection in $\mathbb{P}^n$ given by r equations of degrees l_i, $1 \leq i \leq r$ and denote by W the vector bundle $\oplus_{i=1}^r \mathcal{O}(l_i)$ over $\mathbb{P}^n$. We will be interested in the following cases:

i) $\sum\limits_{i=1}^r l_i < n$;

ii) $\sum\limits_{i=1}^r l_i = n$;

iii) $\sum\limits_{i=1}^r l_i = n + 1$.

Observe that case iii) implies that X is Calabi-Yau. It is useful for our purpose to recall that the only Calabi-Yau threefolds are:

- a quintic in $\mathbb{P}^4$,
- the intersection of a quadric and a quartic in $\mathbb{P}^5$,
- the intersection of two cubics in $\mathbb{P}^5$,
- the intersection of two quadrics and a cubic in $\mathbb{P}^6$.

In each case we can consider a differential equation as explained in the Introduction. We are going to prove how solutions of this equation can be recovered from functions s_β introduced in the previous section, i.e. from "manipulated" solutions of the equation $\nabla_\hbar s = 0$. To this extent, we need to find an explicit form of a suitable class $S(t, \hbar)$ in $H^*(\mathbb{P}^n)$: we will then make use of equivariant cohomology theory and graph theory.

4.1. FIXED POINTS ON THE MODULI SPACES OF STABLE MAPS.

The action of the $(n + 1)-$ dimensional torus T^{n+1} on $\mathbb{P}^n$ induces an action of T^{n+1} on $\overline{\mathcal{M}}_{0,2}(\mathbb{P}^n, d)$. We will describe the set of fixed points. Denote by p_i, $0 \leq i \leq 0$, the fixed points of T^{n+1} acting on $\mathbb{P}^n$: p_i is the projectivization of the $i - th$ coordinate line in $\mathbb{C}^{n+1}$. Also denote $l_{ij} = l_{ji}$, $i \neq j$, the line in $\mathbb{P}^n$ through p_i and p_j. Finally, we call Φ the configuration of fixed points p_i and lines l_{ij}.

Let $[C, x_0, x_1; f]$, with $C = \bigcup_\alpha C_\alpha$, be a fixed point in $Y_{2,d}$. By the definition of the action on $Y_{2,d}$, the geometric image $f(C)$ should be contained in Φ, so it is a union of lines l_{ij}. Obviously, the images of all special points, as being in finite number, are points p_i. Suppose now C_α has less than two special points. Because of the stability of f, this means that $f_\alpha = f_{|C_\alpha} : C_\alpha \to l_{ij} = \mathbb{P}^1$ is a degree d_α covering. When $d_\alpha > 1$, notice that the only branch points of f_α are p_i and p_j, since an element $t \in T^{n+1}$ sends the fiber over a point $q \notin \{p_i, p_j\}$ in the fiber over the point $t \cdot q$. On the other hand, since the fiber of f_α over a fixed point is an orbit with respect to the action of the torus, f_α is a totally ramified covering with two ramification points.

If we introduce homogeneous coordinates z_1, z_2 on C_α, then we have the following diagram

$$
\begin{array}{ccc}
[z_1, z_2] & \longrightarrow & [t_i^{\frac{1}{d_\alpha}} z_1, t_j^{\frac{1}{d_\alpha}} z_2] \\
\downarrow & & \downarrow \\
[0, \dots, X_i, \dots, X_j, \dots, 0] & \longrightarrow & [0, \dots, t_i X_i, \dots, t_j X_j, \dots, 0]
\end{array}
$$

with $X_i = z_1^{d_\alpha}$ and $X_j = z_2^{d_\alpha}$.

By what we have just said, it is also clear that irreducible components with more than two special points are mapped to fixed points in $\mathbb{P}^n$.

We now describe components of fixed points in terms of a special class of graphs. In the sequel, we shall consider the family $\mathcal{G}$ of finite graphs Γ with V_Γ and E_Γ respectively its set of vertices and edges. Each graph Γ is also equipped with three additional data:

- a map $\delta : V_\Gamma \to \{0, \dots, n\}$,
- a map $\lambda : E_\Gamma \to \mathbb{Z}^+$,
- a map $P : V \to \mathcal{P}(\{0,1\})$ such that $\{0,1\}$ is the disjoint union of the $P(v)$, for $v \in V_\Gamma$.

We set $h_v = \#P(v)$, and denote by l_v the valency of v, that is, the cardinality of the set L_v of edges issuing from v. Finally, we say that two graphs Γ_1 and Γ_2 are isomorphic if there exist bijections between V_{Γ_1} and V_{Γ_2}, E_{Γ_1} and E_{Γ_2} which preserve adjacency and additional data. In the sequel, we shall denote by $[\Gamma]$ an equivalence class of graphs in $\mathcal{G}$.

We now construct a graph Γ associated to a fixed point $[C, x_0, x_1; f]$ (see [K]). The vertices $v \in V_\Gamma$ correspond to the connected components C_v of $f^{-1}(p_0, \dots, p_n)$, i.e. points or non-empty union of contracted components of C; the edges $\alpha \in E_\Gamma$ correspond to irreducible components C_α of genus zero mapping to lines l_{ij}. Moreover we define the map P so that it assigns to each vertex the indices of marked points lying in C_v. It is quite obvious that Γ is a connected tree (there are no simple loops for the choice of the set of vertices). Notice that the map δ is well defined!

Alternatively, given an element $[C, x_0, x_1; f] \in \overline{\mathcal{M}}_{0,2}(\mathbb{P}^n, d)^{T^{n+1}}$, we associate to it a graph in $\mathcal{G}$. Each edge α connecting two vertices, with labels i and j, corresponds to an irreducible component mapped to l_{ij} with degree $\lambda(\alpha)$, whereas each vertex v with $\delta(v) = k$ corresponds to connected components C_v mapped to p_k. Finally, if $P(v) \neq \emptyset$, then C_v contains marked points with indices prescribed by $P(v)$. Notice that we don't fix the number of connected components contracted to points. For example, if Γ is the graph of figure 2, then the corresponding element $[C, x_0, x_1; f]$ is a point, since degrees d maps from C to $\mathbb{P}^1$ with two ramification points define the same element in $\overline{\mathcal{M}}_{0,2}(\mathbb{P}^1, d)$.

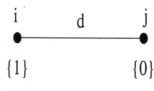

FIGURE 2.

Let Σ_Γ be the locus of fixed points in $\overline{\mathcal{M}}_{0,2}(\mathbb{P}^n, d)$ whose associated graph is Γ. Denote by $\tilde{V}_\Gamma$ the set of vertices for which $h_{v_i} + l_{v_i} - 3 \geq 0$ and choose, for each $v \in \tilde{V}_\Gamma$, an ordering of L_v.

Proposition 4.1.

$$\Sigma_\Gamma \cong \frac{\prod\limits_{i=1}^{s} \overline{\mathcal{M}}_{0, h_{v_i} + l_{v_i}}}{Aut(\Gamma)}.$$

Proof. Consider the morphism

$$\xi_\Gamma : \prod_{i=1}^{s} \overline{\mathcal{M}}_{0, h_{v_i} + l_{v_i}} \longrightarrow \overline{\mathcal{M}}_{0,2}(\mathbb{P}^n, d)$$

which is defined as follows. The image of a point

$$\left([C_1; x_{v_1, 1}, \ldots, x_{v_1, h_{v_1} + l_{v_1}}], \ldots, [C_s; x_{v_s, 1}, \ldots, x_{v_s, h_{v_s} + l_{v_s}}] \right)$$

is an element associated with the graph Γ whose contracted components are given by $C_1, \ldots, C_s$. One can easily verify that Σ_Γ is the image of ξ_Γ and does not depend on the choice of orderings of the L_v, $v \in \tilde{V}_\Gamma$. Since two elements in the same fiber of ξ_Γ differ by an automorphism of Γ, the Proposition is proved.

□

4.2. PROJECTIVE COMPLETE INTERSECTIONS WITH $l_1 + \ldots + l_r < n$.

Let X be a smooth projective complete intersection in $\mathbb{P}^n$ given by r equations of degrees l_i, $1 \leq i \leq r$, with $\sum\limits_{i=1}^{r} l_i < n$, and let $m = n - r$. This is the easier case: in fact, in this section we are going to prove that the functions $s_\beta(t, \hbar) := s_{0\beta}(t, \hbar)$, $\beta = 0, \ldots, m$, (components of vector solutions $(s_{0\beta}(t, \hbar), \ldots s_{m\beta}(t, \hbar))$ of the equation $\nabla_\hbar s = 0$) themselves satisfy the Picard-Fuchs equation for the mirror symmetric family of X. To this purpose, we shall give an explicit form of $s_\beta(t, \hbar)$. Fix a basis $\{T_\beta = P^\beta\}$ of $H^*(\mathbb{P}^n)$ and recall that

$$s_\beta(t, \hbar) = \sum_{d \geq 0} e^{dt} \int_{\overline{\mathcal{M}}_{0,2}(\mathbb{P}^n, d)} (e_0)^* \left(P^\beta e^{\frac{Pt}{\hbar}} \right) \frac{\mathcal{E}_{2,d}}{\hbar + c}, \quad \beta = 0, \ldots, m,$$

with e_0 the evaluation on the first marked point. We can also view $s_\beta(t, \hbar)$ as components of the function

$$S(t, \hbar) = e^{\frac{Pt}{\hbar}} \sum_{d \geq 0} e^{dt} (e_0)_* \left(\frac{\mathcal{E}_{2,d}}{\hbar + c} \right) \in H^*(\mathbb{P}^n), \tag{4.1}$$

since by the integration formula, we obtain

$$s_\beta(t, \hbar) = \int_{\mathbb{P}^n} P^\beta S(t, \hbar). \tag{4.2}$$

For $d = 0$, when $\overline{\mathcal{M}}_{0,2}(\mathbb{P}^n, d)$ is not defined, we let $Euler(\oplus_{i=1}^r \mathcal{O}(l_i))$ play the role of $(e_0)_*(\mathcal{E}_{2,0})$.

Theorem 4.2. *Suppose that* $l_1 + \ldots + l_r < n$. *Then*

$$S(t, \hbar) = e^{\frac{Pt}{\hbar}} \sum_{d \geq 0} e^{dt} \frac{\prod_{m=0}^{dl_1}(l_1 P + m\hbar) \ldots \prod_{m=0}^{dl_r}(l_r P + m\hbar)}{\prod_{m=1}^d (P + m\hbar)^{n+1}}.$$

We will postpone the proof of this Theorem at the end of this section.

Corollary 4.3. *The components* $s_\beta(t, \hbar)$, $\beta = 0, \ldots, m$, *form a basis of solutions of the Picard-Fuchs equation*

$$(\hbar \frac{d}{dt})^{n+1-r} F(t, \hbar) = e^t \prod_{j=1}^r l_j \prod_{m=1}^{l_j - 1} \hbar(l_j \frac{d}{dt} + m) F(t, \hbar)$$

for the mirror symmetric family of X.
    Proof. Set

$$\Gamma(d) = \frac{\prod_{m=0}^{dl_1}(l_1 P + m\hbar) \ldots \prod_{m=0}^{dl_r}(l_r P + m\hbar)}{\prod_{m=1}^d (P + m\hbar)^{n+1}}.$$

Applying differential operators $(\hbar\frac{d}{dt})^{n+1-r}$ and $e^t \prod_{j=1}^r l_j \prod_{m=1}^{l_j-1} \hbar(l_j \frac{d}{dt} + m)$ to $S(t,\hbar)$, we obtain

$$\sum_{d\geq 0}(P + d\hbar)^{n+1-r}\Gamma(d)e^{(\frac{P}{\hbar}+d)t}$$

and

$$\prod_{j=1}^r l_j \sum_{d\geq 0} \prod_{m=1}^{l_j-1} [l_j(P + d\hbar) + m\hbar] e^{(\frac{P}{\hbar}+d+1)t}\Gamma(d).$$

On the other hand we have,

$$[P + (d+1)\hbar]^{n+1-r}\,\Gamma(d+1) = \prod_{j=1}^r l_j \sum_{d\geq 0} \prod_{m=1}^{l_j-1} [l_j(P + d\hbar) + m\hbar]\,\Gamma(d),$$

since

$$[P + (d+1)\hbar]^{n+1-r}\frac{\prod_{m=0}^{(d+1)l_1}(l_1 P + m\hbar)\ldots\prod_{m=0}^{(d+1)l_r}(l_r P + m\hbar)}{\prod_{m=1}^{d+1}(P + m\hbar)^{n+1}} =$$

$$= \Gamma(d)\frac{\prod_{m=1}^{l_1-1}(l_1 P + (dl_1 + m)\hbar)\ldots\prod_{m=1}^{l_r-1}(l_r P + (dl_r + m)\hbar)}{[P + (d+1)\hbar]^r}\prod_{i=1}^r [l_i P + (dl_i + l_i)\hbar] =$$

$$= \prod_{j=1}^r l_j \sum_{d\geq 0} \prod_{m=1}^{l_j-1} [l_j(P + d\hbar) + m\hbar]\,\Gamma(d).$$

$\square$

Corollary 4.4. *If* $\dim X \neq 2$, *the cohomology class* p *of the hyperplane section satisfies the relation*

$$p^{n+1-r} = l_1^{l_1}\ldots l_r^{l_r} q p^{l_1+\ldots+l_r-r}$$

in the quantum cohomology of X.

Proof. It follows easily from observations made in section 2.

$\square$

This corollary is consistent with results of Beauville ([Be, Be]) and Jinzenji ([Ji, Ji]) on quantum cohomology of projective hypersurfaces.

Example 4.5. *If* $l_1 = \ldots = l_r = 1$, *then the relation (cfr. section 1) for the small quantum cohomology of* $\mathbb{P}^{n-r}$ *is*

$$p^{n+1-r} = qp.$$

Givental deduces Theorem 4.2 from its equivariant generalization. To this purpose, let $T = (S^1)^k$ act diagonally on $\mathbb{P}^n$, and $T' = (S^1)^r$ acts trivially on $\mathbb{P}^n$ and diagonally on the fiber of the vector bundle $\oplus_{i=1,\dots,r}\mathcal{O}(l_i)$. Recall that

$$H^*_{T \times T'}(\mathbb{P}^n) = \frac{\mathbb{Q}[p, \lambda_0, \dots, \lambda_k]}{\prod\limits_{i=0}^{k}(p - \lambda_i))} \otimes \mathbb{Q}[\mu_1, \dots, \mu_r]$$

and the equivariant Euler class of the vector bundle $\oplus_{i=1,\dots,r}\mathcal{O}(l_i)$ is equal to $\prod\limits_{i=1}^{r}(l_i p - \mu_i)$. Introduce the equivariant counterpart S' of the class S in the $T \times T'$ equivariant cohomology of $\mathbb{P}^n$. This means that we use the equivariant class p instead of P and replace the Euler classes $\mathcal{E}_{2,d}$ and c by their equivariant partners.

Theorem 4.6. *Let* $S'(t, \hbar) = e^{\frac{pt}{\hbar}} \sum\limits_{d \geq 0} e^{dt}(e_0)_* \left(\frac{\mathcal{E}_{2,d}}{\hbar + c}\right)$. *Suppose that* $l_1 + \dots + l_r < n$.
Then

$$S'(t, h) = e^{\frac{pt}{\hbar}} \sum\limits_{d \geq 0} e^{dt} \frac{\prod_{m=0}^{dl_1}(l_1 p - \mu_1 + m\hbar) \dots \prod_{m=0}^{dl_r}(l_r p - \mu_r + m\hbar)}{\prod_{m=1}^{d}(p - \lambda_0 + m\hbar) \dots \prod_{m=1}^{d}(p - \lambda_n + m\hbar)}.$$

Remark 4.7.

The inclusion of a fiber

$$i : \mathbb{P}^n \hookrightarrow \mathbb{P}^n_{T \times T'}$$

induces a morphism

$$i^* : \begin{array}{ccc} H^*_{T \times T'}(\mathbb{P}^n) & \to & H^*(\mathbb{P}^n) \\ q(p, \lambda, \mu) & \to & q(p, 0, 0) \end{array},$$

which is the cup product with the fundamental class of the fiber. Moreover the commutative diagram

$$\begin{array}{ccc} \mathbb{P}^n & \hookrightarrow & \mathbb{P}^n_{T \times T'} \\ \downarrow \pi_{|\mathbb{P}^n} & & \downarrow \pi \\ \{pt\} & \hookrightarrow & (\mathbb{P}^\infty)^{n+1+r} \end{array}$$

induces a diagram in cohomology:

$$\begin{array}{ccc} H^*_{T \times T'}(\mathbb{P}^n) & \to & H^*(\mathbb{P}^n) \\ \downarrow \pi_* & & \downarrow \pi_* \\ H^*(\mathbb{P}^\infty)^{n+1+r} & \to & H^*(\{pt\}) \end{array}$$

and this implies that if we simply put $\lambda_i = 0$, $1 \leq i \leq n$, $\mu_j = 0$, $1 \leq j \leq r$ we pass from equivariant to non equivariant cohomology, from equivariant to non equivariant integral.

In the proof of Theorem 4.6 we write down all formulas for $r = 1$ (it serves the case when X is a hypersurface in $\mathbb{P}^n$ of degree $l < n$). The proof for $r > 1$ differs only for longer product formulas. So our aim is to prove Theorem 4.6 in the following form

$$S'(t,h) = e^{\frac{pt}{\hbar}} \sum_{d \geq 0} e^{dt} \frac{\prod_{m=0}^{dl}(lp - \mu + m\hbar)}{\prod_{m=1}^{d}(p - \lambda_0 + m\hbar) \dots \prod_{m=1}^{d}(p - \lambda_n + m\hbar)}. \qquad (4.3)$$

From now on, with abuse of notation, we shall consider the ring $H^*_{T^{n+1}}(\mathbb{P}^n)$ tensorized with $\mathbb{Q}(\lambda_0, \dots, \lambda_n)$; this is due to the fact that we need to localize on fixed points components. It becomes a vector space on this field with basis $\phi_i = \frac{\prod_{j \neq i}(p - \lambda_j)}{\prod_{j \neq i}(\lambda_i - \lambda_j)}$, $0 \leq i \leq n$. Moreover

$$\langle \phi_i, \phi_j \rangle = \frac{\delta_{ij}}{\prod_{j \neq i}(\lambda_i - \lambda_j)}.$$

We recall that $W_{2,d}$ is the vector bundle on $\overline{M}_{0,2}(\mathbb{P}^n, d)$ whose fiber over the point $[C, x_0, x_1; f]$ is $H^0(C, f^*(\mathcal{O}(l)))$ and $\mathcal{E}_{2,d}$ its equivariant Euler class, with respect to the given linearization of the action. Besides, we define $W'_{2,d}$ via the following exact sequence of vector bundles

$$0 \longrightarrow W'_{2,d} \longrightarrow W_{2,d} \longrightarrow (e_0)^*(\mathcal{O}(l)) \longrightarrow 0$$

and $\mathcal{E}'_{2,d}$ its Euler class.

Lemma 4.8. *For* $0 \leq i \leq n$

$$\prod_{j \neq i}(\lambda_i - \lambda_j) \left\langle \phi_i, S' \right\rangle = e^{\frac{\lambda_i t}{\hbar}}(l\lambda_i - \mu) \sum_{d \geq 0} e^{dt} \int_{\overline{M}_{0,2}(\mathbb{P}^n, d)} (e_0)^*(\phi_i) \frac{\mathcal{E}'_{2,d}}{\hbar + c}.$$

Proof. Obviously, $\mathcal{E}_{2,d} = \mathcal{E}'_{2,d}(e_0)^*(lp - \mu)$. Since $(e_0)_*(\frac{\mathcal{E}'_d}{\hbar + c}) \in H^*_{T^{n+1}}(\mathbb{P}^n)$ may be written as $\sum_{k=0}^n f_k \phi_k$, for $f \in \mathbb{Q}[p, \lambda_0, \dots, \lambda_n]$, we have

$$\prod_{j \neq i} (\lambda_i - \lambda_j) \langle \phi_i, S' \rangle = \frac{1}{2\pi\sqrt{-1}} \sum_{d \geq 0} e^{dt} \int \frac{e^{\frac{pt}{\hbar}} \phi_i (e_0)_* \left(\frac{\mathcal{E}'_{2,d}(e_0)^*(lp - \mu)}{\hbar + c} \right)}{(p - \lambda_0) \dots (p - \lambda_n)} dp =$$

$$= \frac{1}{2\pi\sqrt{-1}} \sum_{d \geq 0} e^{dt} \int \frac{e^{\frac{pt}{\hbar}} \phi_i (lp - \mu)(e_0)_* \left(\frac{\mathcal{E}'_{2,d}}{\hbar + c} \right)}{(p - \lambda_0) \dots (p - \lambda_n)} dp =$$

$$= \frac{1}{2\pi\sqrt{-1}} \sum_{k=0}^n f_k \phi_{k|p=\lambda_i} \sum_{d \geq 0} e^{dt} e^{\frac{\lambda_i t}{\hbar}} (l\lambda_i - \mu) =$$

$$= e^{\frac{\lambda_i t}{\hbar}} (l\lambda_i - \mu) \sum_{d \geq 0} e^{dt} \frac{1}{2\pi\sqrt{-1}} \int \frac{\phi_i (e_0)_* \left(\frac{\mathcal{E}'_{2,d}}{\hbar + c} \right)}{(p - \lambda_0) \dots (p - \lambda_n)} dp =$$

$$= e^{\frac{\lambda_i t}{\hbar}} (l\lambda_i - \mu) \sum_{d \geq 0} e^{dt} \int_{\overline{\mathcal{M}}_{0,2}(\mathbb{P}^n, d)} (e_0)^* (\phi_i) \frac{\mathcal{E}'_{2,d}}{\hbar + c}.$$

$\square$

For the computation made in Lemma 4.8, we introduce functions

$$Z_i(q, \hbar) := 1 + \sum_{d > 0} q^d \prod_{j \neq i} (\lambda_i - \lambda_j) \int_{\overline{\mathcal{M}}_{0,2}(\mathbb{P}^n, d)} (e_0)^* (\phi_i) \frac{\mathcal{E}'_{2,d}}{\hbar + c}, 0 \leq i \leq n.$$

We will determine explicitly these functions by finding 'linear recursive relations' satisfied by them. These relations stem from the computation of the equivariant integral in 4.2 by means of the formula of integration over connected components of fixed points in $\overline{\mathcal{M}}_{0,2}(\mathbb{P}^n, d)$.
We restrict only to those components of elements $[C, x_0, x_1; f]$ whose marked point x_0 is mapped to the i-th point in $\mathbb{P}^n$: indeed, when computing the integral

$$\int_{\overline{\mathcal{M}}_{0,2}(\mathbb{P}^n, d)} (e_0)^* (\phi_i) \frac{\mathcal{E}'_{2,d}}{\hbar + c}, \tag{4.4}$$

ϕ_i has zero localizations at all the fixed points p_j, $j \neq i$.

Proposition 4.9.

$$Z_i(q, \hbar) := 1 + \sum_{d > 0} \left(\frac{q}{\hbar^{n+1-l}} \right)^d \prod_{j \neq i} (\lambda_i - \lambda_j) \int_{\overline{\mathcal{M}}_{0,2}(\mathbb{P}^n, d)} (e_0)^* (\phi_i) \frac{\mathcal{E}'_{2,d}}{1 + \frac{c}{\hbar}} (-c)^{(n+1-l)d-1}.$$

Proof.   Since $\frac{1}{c+\hbar} = \sum_j \frac{(-c)^j}{\hbar^{j+1}}$ and dim $\overline{\mathcal{M}}_{0,2}(\mathbb{P}^n, d) = n + (n+1)d - 1$, the equivariant integral

$$\int_{\overline{\mathcal{M}}_{0,2}(\mathbb{P}^n, d)} (e_0)^* \, (\phi_i) \mathcal{E}'_{2,d} \, (-c)^j$$

gives non-zero contribution only if

$$n + (n+1)d - 1 \le j + n + \deg(\mathcal{E}'_{2,d}) = j + n + ld - 1$$

or

$$j \ge (n + 1 - l)d - 1.$$

$\square$

Consider now a fixed point curve C whose marked point x_0 is indeed mapped to the i-th fixed point in $\mathbb{P}^n$. There are two cases:

(i) the marked point x_0 is situated on an irreducible component of C mapped with some degree d', $d' \ge 1$ onto the line joining the i-th fixed point with the j-th fixed point in $\mathbb{P}^n$ with $i \ne j$;

(ii) the marked point x_0 is situated on a component of C mapped to the i-th fixed point and carrying two or more other special points.

Proposition 4.10. *A type (ii) fixed point component C in $\overline{\mathcal{M}}_{0,2}(\mathbb{P}^n, d)$ gives zero contribution to the integration formula for*

$$\int_{\overline{\mathcal{M}}_{0,2}(\mathbb{P}^n, d)} (e_0)^* \, (\phi_i) \mathcal{E}'_{2,d} \, (-c)^k, \text{ with } k \ge (n + 1 - l)d - 1.$$

Proof. Let Γ be a graph associated to C and denote by v_0 the vertex having label $\{0\}$. Since C is a type (ii) component, $h_{v_0} + l_{v_0} - 3 \ge 0$. By Proposition 4.1, there exists an isomorphism ξ_Γ between C and $\prod\limits_{\substack{v \in V_\Gamma \\ h_v + l_v - 3 \ge 0}} \frac{\overline{\mathcal{M}}_{0, h_v + l_v}}{Aut(\Gamma)}$, with $Aut(\Gamma)$ the group of automorphism of Γ. Consider the projection

$$\pi_{v_0} : \prod\limits_{\substack{v \in V_\Gamma \\ h_v + l_v - 3 \ge 0}} \frac{\overline{\mathcal{M}}_{0, h_v + l_v}}{Aut(\Gamma)} \longrightarrow \frac{\overline{\mathcal{M}}_{0, h_{v_0} + l_{v_0}}}{Aut(\Gamma)}$$

and denote by c' the Chern class of the $Aut(\Gamma)$-invariant line bundle $\mathcal{L}$ on $\overline{\mathcal{M}}_{0, h_{v_0} + l_{v_0}}$ whose fiber at the point $[C, x_{v_0,1} = x_0, \dots, x_{h_{v_0} + l_{v_0}}]$ is the tangent space to C at x_0. By the definiton of the morphism ξ_Γ, we see that $\pi^*_{v_0}(c') = \xi^*_\Gamma(c)$. Thus if we compute 4.4 over the connected component $\mathcal{C}$, we get

$$\int_C \frac{(e_0)^* (\phi_i)\mathcal{E}'_{2,d}}{\mathcal{E}(\mathcal{N}_C)}(-c)^k = \int_{\Pi_v \overline{\mathcal{M}}_{0,h_v+l_v}} \xi_\Gamma^* \left((-c)^k \left(\frac{(e_0)^* (\phi_i)\mathcal{E}'_{2,d}}{\mathcal{E}(\mathcal{N}_C)} \right) \right) =$$

$$= \frac{1}{|Aut(\Gamma)|} \int_{\overline{\mathcal{M}}_{0,h_{v_0}+l_{v_0}}} (-c')^k (\pi_{v_0})_* \left(\xi_\Gamma^* \left(\frac{(e_0)^* (\phi_i)\mathcal{E}'_{2,d}}{\mathcal{E}(\mathcal{N}_C)} \right) \right)$$

with $\mathcal{E}(\mathcal{N}_C)$ the Euler class of the normal bundle $\mathcal{N}_{\overline{\mathcal{M}}_{0,2}(\mathbb{P}^n,d)/C}$. On the other hand, $\dim \overline{\mathcal{M}}_{0,h_{v_0}+l_{v_0}} = h_{v_0} + l_{v_0} - 3 < (n+1-l)d - 1 \le k$, since $l < n$ and $l_{v_0} \le d$. Since the nilpotency degree of c' should not exceed $h_{v_0} + l_{v_0} - 3$, the Lemma follows. □

Consider now a fixed point $[C, x_0, x_1; f]$ in a type (i) component $\mathcal{K}$. By assumption, C can be regarded as the union of a connected curve C'' and an irreducible component C', which carries the marked point x_0 and maps onto l_{ij}, $i \ne j$, with degree d', $1 \le d' \le d$. Moreover, if f'' denotes the restriction of f to C'', then $[C'', y, x_1; f'']$ $(y = C'' \cap C')$ is obviously a fixed point in $\overline{\mathcal{M}}_{0,2}(\mathbb{P}^n, d-d')$. (Observe that if $d = d'$, then $C'' = \emptyset$.) In addition, the graph G associated with $\mathcal{K}$ contains a subgraph $G'_{d'}$ with one edge α labelled d' and two vertices v_1, v_2 having respectively j and $\{\{0\}, i\}$ as labels. If we remove from G the edge α and the vertex v_2, and add to v_1 the label $\{0\}$, we obtain a graph G'' which corresponds to the connected component in $\overline{\mathcal{M}}_{0,2}(\mathbb{P}^n, d-d')$ containing $[C'', y, x_1; f'']$. Notice that G'' is empty when $d' = d$. In this case $\mathcal{K}$ is a point in $\overline{\mathcal{M}}_{0,2}(\mathbb{P}^n, d)$. See figure 3.

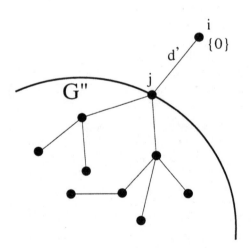

FIGURE 3.

The fundamental idea is to reduce ourselves to integrate on connected component corresponding to "smaller" graphs, i.e., in this case, to write such a formula

$$\int_{K(G)} \ldots = \omega\,(d') \int_{K(G'')} \ldots$$

and from this to build recursive relations among the Z_i. It is easy to check that the automorphism groups of the graphs behave well with respect to this decomposition. Let us first consider the easier case, namely $d = d'$.

Lemma 4.11. *Suppose* $G'' = \emptyset$ *and denote by* $\mathcal{N}_K$ *the normal bundle* $\mathcal{N}_{\overline{\mathcal{M}}_{0,2}(\mathbb{P}^n,d)/K}$. *Then*

$$\prod_{j \neq i}(\lambda_i - \lambda_j) \int_K \frac{e_0^*(\phi_i)\mathcal{E}'_{2,d}(-c)^{(n+1-l)d-1}}{(1 + \frac{c}{\hbar})\mathcal{E}(\mathcal{N}_K)} =$$

$$= \frac{\prod\limits_{m=1}^{ld} [l\lambda_i - \mu + \frac{m}{d}(\lambda_j - \lambda_i)] \left(\frac{\lambda_j - \lambda_i}{d}\right)^{(n+1-l)d-1}}{d\left(1 + \frac{\lambda_i - \lambda_j}{d\hbar}\right) \prod\limits_{\substack{\alpha=0 \\ (\alpha,m)\neq(j,d)}}^{n} \prod\limits_{m=1}^{d} [\lambda_i - \lambda_\alpha + \frac{m}{d}(\lambda_j - \lambda_i)]} := coeff_i^j(d).$$

Proof. Set $K = \{[C', x_0', x_1'; f']\}$. We will prove the Lemma in several steps.

STEP 1) The localization of $e_0^*(\phi_i)$ at K is 1. This follows easily from the following commutative diagram:

$$\begin{array}{ccc} e_0^* : H^*_{T^{n+1}}(\mathbb{P}^n) & \longrightarrow & H^*_{T^{n+1}}(\overline{\mathcal{M}}_{0,2}(\mathbb{P}^n,d)) \\ \delta_{p_i}^* \downarrow & & \downarrow \delta_K^* \\ H^*_{T^{n+1}}(\{p_i\}) & \longrightarrow & H^*_{T^{n+1}}(\{K\}) \end{array}$$

with $\delta_{p_i} : \{p_i\} \to \mathbb{P}^n$ and $\delta_K : \{K\} \to \overline{\mathcal{M}}_{0,2}(\mathbb{P}^n,d)$.

STEP 2) The localization of c at K is $\frac{\lambda_i - \lambda_j}{d}$.
Let $T^{(0)}$ be the equivariant line bundle whose fiber at $[C, x_0, x_1; f]$ is the tangent line to C at x_0. With the same notation adopted in **STEP 1)** we need to compute $c_1(\delta_K^* T^{(0)})$, namely the Chern class of the line bundle

$$T_{x_0'} C' \times_{T^{n+1}} (S^\infty)^{n+1}$$
$$\downarrow$$
$$(\mathbb{P}^\infty)^{n+1} \qquad,$$

with $T_{x_0'} C'$ the tangent line to C' at x_0'. On the other hand, if z_1, z_2 denote homogeneous coordinates on C', the following diagram holds:

$$[z_1, z_2] \qquad \longrightarrow \qquad [t_i^{\frac{1}{d}} z_1, t_j^{\frac{1}{d}} z_2]$$
$$\downarrow \qquad\qquad\qquad\qquad \downarrow$$
$$[0,\dots, z_1^d, z_2^d, \dots, 0] \quad \longrightarrow \quad [0, \dots, t_i z_1^d, t_j z_2^d, \dots, 0].$$

At the point $x_0 = [1,0]$, the action on $T_{x_0'} C'$ is given by

$$\frac{\partial}{\partial z} \mapsto (t_j t_i^{-1})^{\frac{1}{d}} \frac{\partial}{\partial z}, \; z = \frac{z_2}{z_1}.$$

STEP 3) The localization of $\mathcal{E}'_{2,d}$ at $\mathcal{K}$ is $\prod\limits_{m=1}^{ld} [(l\lambda_i - \mu) + \frac{m}{d}(\lambda_j - \lambda_i)]$.

By the same reasoning adopted in **STEP 2)** we need to compute the Euler class of the vector bundle

$$W'_{2,d|_{[C', x_0', x_1' : f']}} \times_{T^{n+1}} (S^\infty)^{n+1} \times_{S^1} (S^\infty)$$
$$\downarrow$$
$$(\mathbb{P}^\infty)^{n+2}.$$

By definition of $W'_{2,d|_{[C', x_0', x_1' : f']}}$, a basis for this fiber is clearly $\{z_2^m z_1^{dl-m}\}_{k=1}^{dl}$. Since the group $T' = S^1$ acts on global sections of $f'^*(\mathcal{O}(l))$ by multiplication by t'^{-1}, the action on the basis is

$$z_2^m z_1^{dl-m} \longrightarrow t'^{-1} t_i^{\left(l - \frac{m}{d}\right)} t_j^{\frac{m}{d}} z_2^m z_1^{dl-m}, \; m \in \{1, \dots, ld\}.$$

Hence the Euler class is

$$\prod_{m=1}^{ld} [(l\lambda_i - \mu) + \frac{m}{d}(\lambda_j - \lambda_i)].$$

STEP 4) The localization of $\mathcal{E}(\mathcal{N}_\mathcal{K})$ at $\mathcal{K}$ is

$$\prod_{j \neq i}^{n} (\lambda_i - \lambda_j) \prod_{\substack{\alpha=0 \\ (\alpha,m) \neq (j,d)}}^{n} \prod_{m=1}^{d} \left(\lambda_i - \lambda_\alpha + \frac{m}{d}(\lambda_j - \lambda_i)\right).$$

Remind that $\mathcal{K}$ is a point and thus the normal bundle is the tangent space to $\overline{\mathcal{M}}_{0,2}(\mathbb{P}^n, d)$ at $\mathcal{K}$. We next define the vector bundle $\mathcal{W}_{f'}$ via the exact sequence

$$0 \to T_{C'} \otimes \mathcal{O}(-x_0' - x_1') \to f'^* T_{\mathbb{P}^n} \to \mathcal{W}_{f'} \to 0.$$

By Remark 3.9, the tangent space to $\overline{\mathcal{M}}_{0,2}(\mathbb{P}^n, d)$ at $\mathcal{K}$ is exactly $H^0(C', \mathcal{W}_{f'})$. Hence,

$$\mathcal{E}(\mathcal{N}_K) = \frac{\mathcal{E}(H^0(C', f'^*T_{\mathbb{P}^n}))}{\mathcal{E}(H^0(C', T_{C'} \otimes \mathcal{O}_{C'}(-x_0' - x_1')))}.$$

We need to compute a basis for $H^0(C', f'^*T_{\mathbb{P}^n})$. Since $f' : C' \to l_{ij} \subseteq \mathbb{P}^n$, the following exact sequence holds

$$0 \to T_{l_{ij}} \to T_{\mathbb{P}^n}|_{l_{ij}} \to \mathcal{N}_{l_{ij}} \to 0.$$

As $\mathbb{P}^n$ is convex, this implies

$$0 \to H^0(C', f'^*T_{l_{ij}}) \to H^0(C', f'^*T_{\mathbb{P}^n}|_{l_{ij}}) \to H^0(C', f'^*\mathcal{N}_{l_{ij}}) \to 0.$$

A basis for $H^0(C', f'^*\mathcal{N}_{l_{ij}})$ is given by $z_1^c z_2^m f'^* \left(\frac{\partial}{\partial X_\alpha} \right)$, with $\alpha \in \{0, \ldots \widehat{i}, \widehat{j}, \ldots, n\}$

and $c + m = d$, $m, c \ge 0$, since $f'^*\mathcal{N}_{l_{ij}} = \overbrace{\mathcal{O}_{C'}(d) \oplus \ldots \oplus \mathcal{O}_{C'}(d)}^{(n-1)-times}$.

On the other hand, a basis for $H^0(C', f'^*T_{l_{ij}})$ is $\left\{ \left(\frac{z_2}{z_1} \right)^a z_2 \frac{\partial}{\partial z_2} \right\}_{a=-d}^{d}$, since $f^*T_{l_{ij}} = \frac{\mathcal{O}_{C'}(d) \oplus \mathcal{O}_{C'}(d)}{\mathbb{C}}$.

Finally, we observe that a basis for $H^0(C', T_{C'} \otimes \mathcal{O}_{C'}(-x_0' - x_1'))$ is $z_2 \frac{\partial}{\partial z_2}$. At this point, the action of the torus T^{n+1} on $H^0(C', f'^*T_{\mathbb{P}^n}|_{l_{ij}})$ is

$$z_1^c z_2^m f'^* \left(\frac{\partial}{\partial X_\alpha} \right) \to t_i^{\frac{a}{d}} t_j^{\frac{m}{d}} t_k^{-1} z_1^c z_2^m f'^* \left(\frac{\partial}{\partial X_\alpha} \right), \quad \alpha \in \{0, \ldots \widehat{i}, \widehat{j}, \ldots, n\}$$

and $c + m = d$, $c, m \ge 0$,

$$\left(\frac{z_2}{z_1} \right)^a z_2 \frac{\partial}{\partial z_2} \to t_j^{\frac{a}{d}} t_i^{-\frac{a}{d}} t_j^{\frac{a}{d}} t_j^{-\frac{a}{d}} \left(\frac{z_2}{z_1} \right)^a z_2 \frac{\partial}{\partial z_2}.$$

In other words, we have

$$\mathcal{E}(\mathcal{N}_K) = \prod_{\substack{a=-d \\ a \ne 0}}^{d} \left(\frac{a(\lambda_j - \lambda_i)}{d} \right) \prod_{\substack{\alpha \ne i \\ \alpha \ne j}}^{n} \prod_{\substack{c+m=d \\ m,c \ge 0}} [\frac{c}{d}\lambda_i + \frac{m}{d}\lambda_j - \lambda_\alpha] =$$

$$= \prod_{\alpha \ne \{i,j\}} (\lambda_i - \lambda_\alpha) \prod_{\substack{a=-d \\ a \ne 0}}^{d} \left(\frac{a(\lambda_j - \lambda_i)}{d} \right) \prod_{\substack{\alpha \ne i \\ \alpha \ne j}}^{n} \prod_{m=1}^{d} [\lambda_i - \lambda_\alpha + \frac{m}{d}(\lambda_j - \lambda_i)] =$$

$$= \prod_{\alpha \ne \{i,j\}} (\lambda_i - \lambda_\alpha) \prod_{\substack{a=-d \\ a \ne 0}}^{-1} \left(\frac{a(\lambda_j - \lambda_i)}{d} \right) \prod_{\alpha \ne j}^{n} \prod_{m=1}^{d} [\lambda_i - \lambda_\alpha + \frac{m}{d}(\lambda_j - \lambda_i)] =$$

$$= \prod_{\alpha \neq \{i,j\}} (\lambda_i - \lambda_\alpha)(\lambda_i - \lambda_j) \prod_{\substack{\alpha=0 \\ (\alpha,m)\neq(j,d)}}^{n} \prod_{m=1}^{d} [\lambda_i - \lambda_\alpha + \frac{m}{d}(\lambda_j - \lambda_i)] =$$

$$= \prod_{j \neq i} (\lambda_i - \lambda_j) \prod_{\substack{\alpha=0 \\ (\alpha,m)\neq(j,d)}}^{n} \prod_{m=1}^{d} \left(\lambda_i - \lambda_\alpha + \frac{m}{d}(\lambda_j - \lambda_i) \right).$$

To summarize, we have

$$\prod_{j \neq i} (\lambda_i - \lambda_j) \int_{\mathcal{K}} \frac{e_0^*(\phi_i) \mathcal{E}'_{2,d}(-c)^{(n+1-l)d-1}}{(1 + \frac{c}{\hbar}) \mathcal{E}(\mathcal{N}_\mathcal{K})} =$$

$$= \frac{\prod_{m=1}^{ld} [l\lambda_i - \mu + \frac{m}{d}(\lambda_j - \lambda_i)] \left(\frac{\lambda_j - \lambda_i}{d} \right)^{(n+1-l)d-1}}{\left(1 + \frac{\lambda_i - \lambda_j}{d\hbar}\right) \prod_{\substack{\alpha=0 \\ (\alpha,m)\neq(j,d)}}^{n} \prod_{m=1}^{d} [\lambda_i - \lambda_\alpha + \frac{m}{d}(\lambda_j - \lambda_i)]} \int_{\mathcal{K}} 1,$$

and by the orbifold structure of $\overline{\mathcal{M}}_{0,2}(\mathbb{P}^n, d)$ this equals

$$\frac{\prod_{m=1}^{ld} [l\lambda_i - \mu + \frac{m}{d}(\lambda_j - \lambda_i)] \left(\frac{\lambda_j - \lambda_i}{d} \right)^{(n+1-l)d-1}}{\left(1 + \frac{\lambda_i - \lambda_j}{d\hbar}\right) \prod_{\substack{\alpha=0 \\ (\alpha,m)\neq(j,d)}}^{n} \prod_{m=1}^{d} [\lambda_i - \lambda_\alpha + \frac{m}{d}(\lambda_j - \lambda_i)]} \frac{1}{\#Aut(f' : C' \to l_{ij})} =$$

$$= \frac{\prod_{m=1}^{ld} [l\lambda_i - \mu + \frac{m}{d}(\lambda_j - \lambda_i)] \left(\frac{\lambda_j - \lambda_i}{d} \right)^{(n+1-l)d-1}}{\left(1 + \frac{\lambda_i - \lambda_j}{d\hbar}\right) \prod_{\substack{\alpha=0 \\ (\alpha,m)\neq(j,d)}}^{n} \prod_{m=1}^{d} [\lambda_i - \lambda_\alpha + \frac{m}{d}(\lambda_j - \lambda_i)]} \frac{1}{d}.$$

So the Lemma is proved.

□

Consider now a connected component whose associated graph is the union of two subgraphs as described before Lemma 4.11. Fix d', $1 \leq d' \leq d$ and j (the label of the vertex v_1). Denote by $\mathcal{K}_{G'',j}$ the connected component in $\overline{\mathcal{M}}_{0,2}(\mathbb{P}^n, d - d')$ associated with G''. We can now prove

Lemma 4.12. *For a connected component* $\mathcal{K}$ *as above,*

$$\prod_{j\neq i}(\lambda_i - \lambda_j)\int_{\mathcal{K}} \frac{e_0^*(\phi_i)\mathcal{E}_{2,d}'(-c)^{(n+1-l)d-1}}{(1+\frac{c}{\hbar})\mathcal{E}(\mathcal{N}_{\mathcal{K}})} =$$

$$= \frac{\prod\limits_{m=1}^{ld'}[l\lambda_i - \mu + \frac{m}{d'}(\lambda_j - \lambda_i)]\left(\frac{\lambda_j-\lambda_i}{d'}\right)^{(n+1-l)d-1}}{d'\left(1+\frac{\lambda_i-\lambda_j}{d'\hbar}\right)\prod\limits_{\substack{\alpha=0\\(\alpha,m)\neq(j,d')}}^{n}\prod\limits_{m=1}^{d'}[\lambda_i - \lambda_\alpha + \frac{m}{d'}(\lambda_j - \lambda_i)]}\cdot$$

$$\cdot\prod_{k\neq j}(\lambda_j - \lambda_k)\int_{\mathcal{K}_{G'',j}}\frac{\mathcal{E}_{2,d-d'}'e_0^*(\phi_j)}{\mathcal{E}(\mathcal{N}_{\mathcal{K}_{G'',j}})(\frac{\lambda_j-\lambda_i}{d'}+c)} =$$

$$= \left(\frac{\lambda_j-\lambda_i}{d'}\right)^{(n+1-l)(d-d')}coeff_i^j(d')\prod_{k\neq j}(\lambda_j - \lambda_k)\int_{\mathcal{K}_{G'',j}}\frac{\mathcal{E}_{2,d-d'}'e_0^*(\phi_j)}{\mathcal{E}(\mathcal{N}_{\mathcal{K}_{G'',j}})(\frac{\lambda_j-\lambda_i}{d'}+c)}.$$

Proof. By our assumptions, we can regard $\mathcal{K}$ in the following way. Consider $\overline{\mathcal{M}}_{0,2}(\mathbb{P}^n, d')$ and $\overline{\mathcal{M}}_{0,2}(\mathbb{P}^n, d-d')$, and denote by e_0'' and e_1' the evaluation maps

$$e_1' : \overline{\mathcal{M}}_{0,2}(\mathbb{P}^n, d') \longrightarrow \mathbb{P}^n,$$

$$e_0'' : \overline{\mathcal{M}}_{0,2}(\mathbb{P}^n, d-d') \longrightarrow \mathbb{P}^n.$$

Next let $\mathcal{K}_{d',j}$ the connected component of fixed points in $\overline{\mathcal{M}}_{0,2}(\mathbb{P}^n, d')$ associated with the graph of figure 4, and $\mathcal{K}_{G'',j}$ the connected component in $\overline{\mathcal{M}}_{0,2}(\mathbb{P}^n, d-d')$ associated with the graph G''. Then $\mathcal{K} \simeq \mathcal{K}_{d',j} \times \mathcal{K}_{G'',j} \subseteq \overline{\mathcal{M}}_{0,2}(\mathbb{P}^n, d') \times_{\mathbb{P}^n} \overline{\mathcal{M}}_{0,2}(\mathbb{P}^n, d-d')$, where the fiber product is defined via the evaluation maps e_0'' and e_1'. Even in this Lemma, we proceed in several steps.

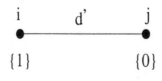

FIGURE 4.

STEP 1) The localization of $e_0^*(\phi_i)$ at $\mathcal{K}$ is 1. This is similar to **STEP 1)** of Lemma 4.11.

STEP 2) Since the line bundle $T^{(0)}$ is trivial on $\mathcal{K}$, the localization of c is computed in the same manner as in **STEP 2)** of Lemma 4.11.

STEP 3) Consider the projections

$$\pi_1 : \mathcal{K} \longrightarrow \mathcal{K}_{d',j},$$

$$\pi_2 : \mathcal{K} \longrightarrow \mathcal{K}_{G'',j},$$

and define the vector bundles

$$\mathcal{F}_{2,d'} := \pi_1^*(W_{2,d'}) \text{ and}$$

$$\mathcal{F}_{2,d-d'} := \pi_2^*(W_{2,d-d'}).$$

If

$$\nu : \mathcal{K} \longrightarrow \mathbb{P}^n$$

is the evaluation map on the point corresponding to the vertex v_1, then we have the following exact sequence

$$0 \to W_{2,d} \to \mathcal{F}_{2,d'} \oplus \mathcal{F}_{2,d-d'} \to \nu^*(\mathcal{O}(l)) \to 0.$$

Moreover,

$$0 \to W'_{2,d} \to W_{2,d} \to e_0^*(\mathcal{O}(l)) \to 0,$$

$$0 \to W'_{2,d'} \to W_{2,d'} \to e_0'^*(\mathcal{O}(l)) \to 0,$$

$$0 \to W'_{2,d-d'} \to W_{2,d-d'} \to e_0''^*(\mathcal{O}(l)) \to 0.$$

Since $e_0' \circ \pi_1 = e_0$ and $e_0'' \circ \pi_2 = \nu$, the Euler class of $W'_{2,d}$ is

$$\mathcal{E}'_{2,d} = \mathcal{E}'_{2,d'}\mathcal{E}'_{2,d-d'} \frac{\mathcal{E}(\pi_1^* e_0'^*(\mathcal{O}(l)))\mathcal{E}(\pi_2^* e_0''^*(\mathcal{O}(l)))}{\mathcal{E}(e_0^*(\mathcal{O}(l)))\mathcal{E}(\nu^*(\mathcal{O}(l)))} = \mathcal{E}'_{2,d'}\mathcal{E}'_{2,d-d'}.$$

Observe that the localization of $\mathcal{E}'_{2,d'}$ is computed in **STEP 3)** of Lemma 4.11.

STEP 4) By our hypotheses on elements $[C, x_0, x_1, f]$ of $\mathcal{K}$, $C = C' \cup C''$, with C' mapped to l_{ij} with degree d'. Denote by p (mapped to p_j in $\mathbb{P}^n$) the intersection point of C' and C'' and by $\{q_1, \ldots, q_s\}$ the remaining nodes of C. We define the vector bundle $\mathcal{L}$ via the exact sequence

$$0 \to \mathcal{N}_\mathcal{K} \to \pi_1^*(\mathcal{N}_{\mathcal{K}_{j,d'}}) \oplus \pi_2^*(\mathcal{N}_{\mathcal{K}_{G'',j}}) \to \mathcal{L} \to 0.$$

We also have

$$0 \to H^0(C, f^* T_{\mathbb{P}^n}) \to H^0(C', f'^* T_{\mathbb{P}^n}) \oplus H^0(C'', f''^* T_{\mathbb{P}^n}) \to f^* T_{p_j} \to 0,$$

$$0 \to H^0(C', T_{C'}(-p)) \to H^0(C', T_{C'}) \to T_p \to 0,$$

$$0 \to H^0\left(C', T_{C'}(-p - \sum_{t=1}^{s} q_t)\right) \to H^0\left(C'', T_{C''}(-\sum_{t=1}^{s} q_t)\right) \to T_p \to 0,$$

where $f' = f_{|_{C'}}$ and $f'' = f_{|_{C''}}$ and, by abuse of notation, we still denote by p the points on C' and C'' which correspond to $p = C' \cap C''$.

Moreover, the deformations which preserve the combinatorial type of C are exactly the deformations which preserve the combinatorial type of C' and those of C''. Keeping in mind Remark 3.9, the Euler class of $\mathcal{L}$ is

$$\frac{\mathcal{E}(T_p C' \otimes T_p C'')}{\mathcal{E}(f^* T_{p_j})}.$$

Let c denote again the Euler class of $T_p C''$ and observe that $\mathcal{E}(T_p C')$ is given by computing the character of T^{n+1} on the tangent space to C' at p, since $T_p C'$ is trivial on $\mathcal{K}_{j,d'}$. If we introduce homogeneous coordinates z_1, z_2 on C', we remind that the action of T^{n+1} is

$$[z_1, z_2] \longrightarrow [t_i^{\frac{1}{d'}} z_1, t_j^{\frac{1}{d'}} z_2]$$

and if $p = [0, 1]$, then we have

$$\frac{z_1}{z_2} \to t_i^{\frac{1}{d'}} t_j^{-\frac{1}{d'}} \frac{z_1}{z_2}.$$

Hence $\mathcal{E}(T_p C') = \frac{\lambda_j - \lambda_i}{d'}$. Finally, $\mathcal{E}(f^* T_{p_j}) = \prod\limits_{k \neq j} (\lambda_j - \lambda_k)$, since a basis for $f^* T_{p_j}$ is given $\frac{\partial}{\partial X_k}$, $k \neq j$. We have come to the conclusion

$$\mathcal{E}(\mathcal{N}_\mathcal{K}) = \frac{\mathcal{E}(\mathcal{N}_{\mathcal{K}_{G'',j}}) \frac{\lambda_j - \lambda_i}{d'} \prod\limits_{j \neq i} (\lambda_i - \lambda_j) \prod\limits_{\substack{\alpha=0 \\ (\alpha,m) \neq (j,d)}}^{n} \prod\limits_{m=1}^{d} (\lambda_i - \lambda_\alpha + \frac{m}{d}(\lambda_j - \lambda_i))}{\prod\limits_{k \neq j} (\lambda_j - \lambda_k)},$$

because of **STEP 4)** of Lemma 4.11.

To summarize, we have

$$\prod_{j \neq i}(\lambda_i - \lambda_j) \int_K \frac{e_0^*(\phi_i)\mathcal{E}_{2,d}'(-c)^{(n+1-l)d-1}}{(1 + \frac{c}{\hbar})\mathcal{E}(\mathcal{N}_K)} =$$

$$= \frac{\prod_{m=1}^{ld'} [l\lambda_i - \mu + \frac{m}{d'}(\lambda_j - \lambda_i)]\left(\frac{\lambda_j - \lambda_i}{d'}\right)^{(n+1-l)d-1}}{\left(1 + \frac{\lambda_i - \lambda_j}{d'\hbar}\right)\prod_{\substack{\alpha=0 \\ (\alpha,m)\neq(j,d')}}^{n}\prod_{m=1}^{d'}[\lambda_i - \lambda_\alpha + \frac{m}{d'}(\lambda_j - \lambda_i)]} \cdot$$

$$\cdot \prod_{k \neq j}(\lambda_j - \lambda_k) \int_{K_{G'',j}} \frac{\mathcal{E}_{2,d-d'}' e_0^*(\phi_j)}{\mathcal{E}(\mathcal{N}_{K_{G'',j}})(\frac{\lambda_j - \lambda_i}{d'} + c)} \int_{K_{d',j}} 1 =$$

$$= \frac{\prod_{m=1}^{ld'} [l\lambda' - \mu + \frac{m}{d'}(\lambda_j - \lambda_i)]\left(\frac{\lambda_j - \lambda_i}{d'}\right)^{(n+1-l)d-1}}{d'\left(1 + \frac{\lambda_i - \lambda_j}{d'\hbar}\right)\prod_{\substack{\alpha=0 \\ (\alpha,m)\neq(j,d')}}^{n}\prod_{m=1}^{d'}[\lambda_i - \lambda_\alpha + \frac{m}{d'}(\lambda_j - \lambda_i)]} \cdot$$

$$\cdot \prod_{k \neq j}(\lambda_j - \lambda_k) \int_{K_{G'',j}} \frac{\mathcal{E}_{2,d-d'}' e_0^*(\phi_j)}{\mathcal{E}(\mathcal{N}_{K_{G'',j}})(\frac{\lambda_j - \lambda_i}{d'} + c)}.$$

$\square$

Theorem 4.13. *Set* $z_i(Q, \hbar) = Z_i(\hbar^{n+1-l}Q, \hbar)$. *Then*

$$z_i(Q, \hbar) = 1 + \sum_{d'>0} Q^{d'} \sum_{j \neq i} coeff_i^j(d') z_j(Q, \frac{\lambda_j - \lambda_i}{d'}),$$

where

$$coeff_i^j(d') = \frac{\prod_{m=1}^{ld'} [l\lambda_i - \mu + \frac{m}{d'}(\lambda_j - \lambda_i)]\left(\frac{\lambda_j - \lambda_i}{d'}\right)^{(n+1-l)d'-1}}{d'\left(1 + \frac{\lambda_i - \lambda_j}{d'\hbar}\right)\prod_{\substack{\alpha=0 \\ (\alpha,m)\neq(j,d')}}^{n}\prod_{m=1}^{d'}[\lambda_i - \lambda_\alpha + \frac{m}{d'}(\lambda_j - \lambda_i)]}.$$

Proof. By the integration formula over connected components $\mathcal{K}$ of fixed points, we obtain

$$z_i(Q,\hbar) = 1 + \sum_{d>0} Q^d \prod_{j\neq i}(\lambda_i - \lambda_j) \int_{\overline{\mathcal{M}}_{0,2}(\mathbb{P}^n,d)} (e_0)^*(\phi_i)\frac{\mathcal{E}'_{2,d}}{1+\frac{c}{\hbar}}(-c)^{(n+1-l)d-1} =$$

$$= 1 + \sum_{d>0} Q^d \sum_{\mathcal{K}} \prod_{j\neq i}(\lambda_i - \lambda_j) \int_{\mathcal{K}} (e_0)^*(\phi_i)\frac{\mathcal{E}'_{2,d}}{(1+\frac{c}{\hbar})\,\mathcal{E}(\mathcal{N}_\mathcal{K})}(-c)^{(n+1-l)d-1} =$$

$$= 1 + \sum_{d>0} Q^d \sum_{j\neq i} \sum_{d'=1}^{d} \sum_{\mathcal{K}_{j,d'}} \prod_{j\neq i}(\lambda_i - \lambda_j)\cdot$$

$$\cdot \int_{\mathcal{K}_{j,d'}} (e_0)^*(\phi_i)\frac{\mathcal{E}'_{2,d}}{(1+\frac{c}{\hbar})\,\mathcal{E}(\mathcal{N}_{\mathcal{K}_{j,d'}})}(-c)^{(n+1-l)d-1},$$

where $\mathcal{K}_{j,d'}$ is the connected component associated with the graph of figure 5.

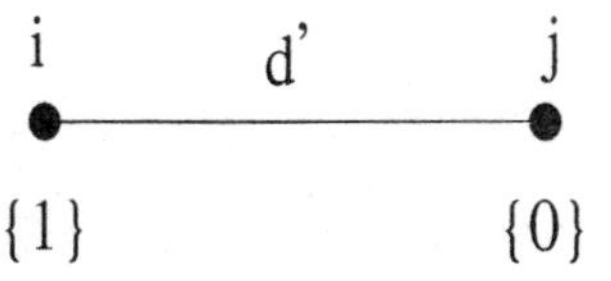

FIGURE 5.

Therefore, by Lemma 4.11 and Lemma 4.12 we have

$$z_i(Q,\hbar) = 1 + \sum_{d>0} Q^d \sum_{j\neq i} \left\{ \sum_{d'=1}^{d-1} \left(\frac{\lambda_j - \lambda_i}{d'}\right)^{(n+1-l)(d-d')} \quad coeff_i^j(d') \prod_{k\neq j}(\lambda_j - \lambda_k)\cdot \right.$$

$$\left. \cdot \sum_{\mathcal{K}_{G'',j}} \int_{\mathcal{K}_{G'',j}} \frac{\mathcal{E}'_{2,d-d'}\,e_0^*(\phi_j)}{\mathcal{E}(\mathcal{N}_{\mathcal{K}_{G'',j}})(\frac{\lambda_j-\lambda_i}{d'}+c)} + coeff_i^j(d) \right\} =$$

$$= 1 + \sum_{d'+(d-d')>0} \sum_{j\neq i} \left\{ \sum_{d'=1}^{d-1} Q^{d'} \left(\frac{\lambda_j - \lambda_i}{d'}\right)^{(n+1-l)(d-d')} \quad coeff_i^j(d') \prod_{k\neq j}(\lambda_j - \lambda_k)\cdot \right.$$

$$\cdot \sum_{\mathcal{K}_{G'',j}} Q^{d-d'} \int_{\mathcal{K}_{G'',j}} \frac{\mathcal{E}'_{2,d-d'} e_0^*(\phi_j)}{\mathcal{E}(\mathcal{N}_{\mathcal{K}_{G'',j}})(\frac{\lambda_j - \lambda_i}{d'} + c)} + Q^d coeff_i^j(d) \Bigg\} =$$

$$= 1 + \sum_{d'+(d-d')>0} \sum_{j\neq i} \Bigg\{ \sum_{d'=1}^{d-1} Q^{d'} coeff_i^j(d') \prod_{k\neq j}(\lambda_j - \lambda_k) \cdot$$

$$\cdot \sum_{\mathcal{K}_{G'',j}} Q^{d-d'} \int_{\mathcal{K}_{G'',j}} \frac{\mathcal{E}'_{2,d-d'} e_0^*(\phi_j)(-c)^{(n+1-l)(d-d')-1}}{\mathcal{E}(\mathcal{N}_{\mathcal{K}_{G'',j}})(1 + \frac{cd'}{\lambda_j - \lambda_i})} + Q^d coeff_i^j(d) \Bigg\} =$$

$$= 1 + \sum_{d'>0} Q^{d'} \sum_{j\neq i} coeff_i^j(d') \Bigg\{ 1 + \sum_{d''>0} Q^{d''} \prod_{k\neq j}(\lambda_j - \lambda_k) \cdot$$

$$\cdot \sum_{\mathcal{K}_{G'',j}} \int_{\mathcal{K}_{G'',j}} (e_0)^*(\phi_j) \frac{\mathcal{E}'_{2,d''}}{\mathcal{E}(\mathcal{N}_{\mathcal{K}_{G'',j}})(1 + \frac{cd'}{\lambda_j - \lambda_i})} (-c)^{(n+1-l)d''-1} \Bigg\} =$$

$$= 1 + \sum_{d'>0} Q^{d'} \sum_{j\neq i} coeff_i^j(d') z_j(Q, \frac{\lambda_j - \lambda_i}{d'}).$$

□

We now determine explicitly the functions $z_i(Q, \hbar)$.

Proposition 4.14. *The functions $z_i(Q, 1/\omega), 0 \leq i \leq n, \omega = 1/\hbar$, are power series $\sum_{d\geq 0} C_i(d, 1/\omega)Q^d$ in Q with coefficients $C_i(d, 1/\omega)$ which are rational functions of ω with poles of order at least one at $\omega = \frac{d'}{\lambda_j - \lambda_i}$, with $d' = 1, \ldots, d$. The functions z_i are uniquely determined by these properties, the recursive relations of Theorem 4.13 and the initial condition $C_i(0) = 1$.*

Proof. The proof is immediate from the recursion relations and from the nature of the coefficients, which have poles only at the indicated points.

□

Proposition 4.15. *The series*

$$z_i(Q, 1/\omega) = \sum_{d\geq 0} Q^d \frac{\prod_{m=1}^{ld} [(l\lambda_i - \mu)\omega + m]}{d! \prod_{\alpha\neq i} \prod_{m=1}^{d} [(\lambda_i - \lambda_\alpha)\omega + m]} \tag{4.5}$$

satisfy all the conditions of Proposition 4.14.

Proof. We need to prove that the coefficients of functions in (4.5) satisfy the following recursive relations:

$$C_i(d, 1/\omega) = \sum_{j \neq i} \sum_{a=1}^{d} coeff_i^j(a) C_j(d - a, \frac{\lambda_j - \lambda_i}{a}).$$

To achieve this goal, we decompose the coefficients of the power series in (4.5) into sum of simple fractions, i.e.

$$\frac{\prod_{m=1}^{ld} [(l\lambda_i - \mu)\omega + m]}{d! \prod_{\alpha \neq i} \prod_{m=1}^{d} [(\lambda_i - \lambda_\alpha)\omega + m]} =$$

$$= \sum_{j \neq i} \sum_{a=1}^{d} \frac{1}{[(\lambda_i - \lambda_j)\omega + a]} \frac{\prod_{m=1}^{ld} \left(\frac{l\lambda_i - \mu}{\lambda_j - \lambda_i} a + m\right)}{d! \prod_{\substack{\alpha \neq i \\ (\alpha,m) \neq (j,a)}} \prod_{m=1}^{d} \left(\frac{\lambda_i - \lambda_\alpha}{\lambda_j - \lambda_i} a + m\right)}.$$

On the other hand, we have

$$\sum_{j \neq i} \sum_{a=1}^{d} coeff_i^j(a) C_j(d - a, \frac{\lambda_j - \lambda_i}{a}) =$$

$$= \sum_{j \neq i} \sum_{a=1}^{d} \frac{1}{[(\lambda_i - \lambda_j)\omega + a]} \frac{\prod_{m=1}^{la} \left(\frac{l\lambda_i - \mu}{\lambda_j - \lambda_i} a + m\right)}{\prod_{\substack{\alpha \\ (\alpha,m) \neq (j,a)}} \prod_{m=1}^{a} \left(\frac{\lambda_i - \lambda_\alpha}{\lambda_j - \lambda_i} a + m\right)} \cdot$$

$$\cdot \frac{\prod_{m=1}^{l(d-a)} \left(\frac{l\lambda_j - \mu}{\lambda_j - \lambda_i} a + m\right)}{(d - a)! \prod_{\alpha \neq j} \prod_{m=1}^{d-a} \left(\frac{\lambda_j - \lambda_\alpha}{\lambda_j - \lambda_i} a + m\right)} =$$

$$= \sum_{j \neq i} \sum_{a=1}^{d} \frac{1}{[(\lambda_i - \lambda_j)\omega + a]} \frac{\prod_{m=1}^{la} \left(\frac{l\lambda_i - \mu}{\lambda_j - \lambda_i} a + m\right)}{\prod_{\substack{\alpha \\ (\alpha,m) \neq (j,a)}} \prod_{m=1}^{a} \left(\frac{\lambda_i - \lambda_\alpha}{\lambda_j - \lambda_i} a + m\right)} \frac{\prod_{m=la+1}^{ld} \left(\frac{l\lambda_i - \mu}{\lambda_j - \lambda_i} a + m\right)}{\prod_{\alpha} \prod_{m=a+1}^{d} \left(\frac{\lambda_i - \lambda_\alpha}{\lambda_j - \lambda_i} a + m\right)} =$$

$$= \sum_{j \neq i} \sum_{a=1}^{d} \frac{1}{[(\lambda_i - \lambda_j)\omega + a]} \frac{\prod_{m=1}^{ld} \left(\frac{l\lambda_i - \mu}{\lambda_j - \lambda_i} a + m \right)}{(d-a)! \prod_{\alpha \neq j} \prod_{m=1}^{d} \left(\frac{\lambda_i - \lambda_\alpha}{\lambda_j - \lambda_i} a + m \right) \prod_{m=1}^{a-1} (-a + m)} =$$

$$= \sum_{j \neq i} \sum_{a=1}^{d} \frac{1}{[(\lambda_i - \lambda_j)\omega + a]} \frac{\prod_{m=1}^{ld} \left(\frac{l\lambda_i - \mu}{\lambda_j - \lambda_i} a + m \right)}{d! \prod_{\substack{\alpha \neq i \\ (\alpha,m) \neq (j,a)}} \prod_{m=1}^{d} \left(\frac{\lambda_i - \lambda_\alpha}{\lambda_j - \lambda_i} a + m \right)} = C_i(d, 1/\omega).$$

□

We can finally prove Theorem 4.6 in the form 4.3.

Proof of Theorem 4.6. By Proposition 4.15 and Lemma 4.8,

$$\prod_{j \neq i} (\lambda_i - \lambda_j) \langle \phi_i, S' \rangle = e^{\frac{\lambda_i t}{\hbar}} (l\lambda_i - \mu) \sum_{d \geq 0} \frac{e^{dt}}{\hbar^{(n+1-l)d}} \frac{\prod_{m=1}^{ld} [\frac{(l\lambda_i - \mu)}{\hbar} + m]}{d! \prod_{\alpha \neq i} \prod_{m=1}^{d} [\frac{(\lambda_i - \lambda_\alpha)}{\hbar} + m]},$$

with $\omega = 1/\hbar$. Hence

$$\prod_{j \neq i} (\lambda_i - \lambda_j) \langle \phi_i, S' \rangle = e^{\frac{\lambda_i t}{\hbar}} (l\lambda_i - \mu) \sum_{d \geq 0} \frac{e^{dt}}{\hbar^d} \frac{\prod_{m=1}^{ld} [(l\lambda_i - \mu) + m\hbar]}{d! \prod_{\alpha \neq i} \prod_{m=1}^{d} [(\lambda_i - \lambda_\alpha) + m\hbar]} =$$

$$= \frac{1}{2\pi\sqrt{-1}} \int \phi_i e^{\frac{pt}{\hbar}} \sum_{d \geq 0} \frac{e^{dt}}{\hbar^d} \sum_{d \geq 0} e^{dt} \frac{\prod_{m=1}^{ld} [(lp - \mu) + m\hbar]}{\prod_{\alpha=0}^{n} \prod_{m=1}^{d} [(p - \lambda_\alpha) + m\hbar] \prod_{\alpha=0}^{n} (p - \lambda_\alpha)} dp$$

This implies that

$$S'(t, \hbar) = e^{\frac{pt}{\hbar}} \sum_{d \geq 0} e^{dt} \frac{\prod_{m=1}^{ld} [(lp - \mu) + m\hbar]}{\prod_{\alpha=0}^{n} \prod_{m=1}^{d} [(p - \lambda_\alpha) + m\hbar]}.$$

□

4.3. Projective complete intersections with $l_1 + \ldots + l_r = n$

Let $X \subset \mathbb{P}^n$ be a non-singular complete intersection given by equations of degrees $l_1, \ldots, l_r$ with $l_1 + \ldots + l_r = n$. As in the previous section, we look for recursive relations so as to determine explicitly functions $s_\beta(t, \hbar)$.

However, in this case the recursive relations involve more terms, since we have to modify Proposition 4.10 as explained in Lemma 4.22, that is, it is not any more true that these components give zero contribution.

This will imply that the two generating function, the one which contains numbers counting curves, and the one containing solutions of Picard Fuchs equation, do not coincide exactly, as they did in the previous case, but we can pass from one another with suitable transformations.

Let us now explain in details; with the same notations introduced in the previous section, let $S(t, \hbar) \in H^*(\mathbb{P}^n)$ be defined as in 4.1 and denote by $S'(t, \hbar)$ its equivariant counterpart. The main theorem of this section is

Theorem 4.16. *Suppose $l_1 + \ldots + l_r = n$. Then*

$$S'(t, \hbar) = e^{\frac{pt - l_1! \ldots l_r! e^t}{\hbar}} \sum_{d \geq 0} e^{dt} \frac{\prod_{m=0}^{dl_1} (l_1 p - \mu_1 + m\hbar) \ldots \prod_{m=0}^{dl_r} (l_r p - \mu_r + m\hbar)}{\prod_{m=1}^{d} (p - \lambda_0 + m\hbar) \ldots \prod_{m=1}^{d} (l_1 p - \lambda_n + m\hbar)}.$$

As explained in Remark 4.7, we deduce

Theorem 4.17. *Suppose $l_1 + \ldots + l_r = n$. Then*

$$S(t, \hbar) = e^{\frac{Pt - l_1! \ldots l_r! e^t}{\hbar}} \sum_{d \geq 0} e^{dt} \frac{\prod_{j=1}^{r} \prod_{m=1}^{d} (l_j P + m\hbar)}{\prod_{m=1}^{d} (P + m\hbar)^{n+1}},$$

with P the generator of the cohomology algebra of $\mathbb{P}^n$.

In this case, the functions $s_\beta(t, \hbar) = \int_{\mathbb{P}^n} P^\beta S(t, \hbar)$ do not satisfy the Picard-Fuchs equation for the mirror symmetric family of X. Nevertheless, if we multiply them by the common factor $e^{\frac{l_1! \ldots l_r! e^t}{\hbar}}$, we can prove that

$$s'_{(1)\beta}(t,\hbar) := e^{\frac{l_1!\ldots l_r!e^t}{\hbar}} \int_{\mathbb{P}^n} P^\beta S(t,\hbar) = e^{\frac{l_1!\ldots l_r!e^t}{\hbar}} s_\beta(t,\hbar), \ \beta = 0,\ldots,m,$$

satisfy the Picard - Fuchs differential equation

$$\left[\left(\hbar\frac{d}{dt} \right)^{n+1-r} - e^t \prod_{j=1}^{r} l_j \prod_{m=1}^{l_j-1} \left(l_j\hbar\frac{d}{dt} + \hbar m \right) \right] F(t,\hbar) = 0.$$

Proof. It is exactly the same computation as in 4.3.

$\square$

Anyway, instead of modifying the functions, we prefer to modify the differential operator, so to obtain information on $SQH^*(X)$.

Proposition 4.18. *The functions $s_\beta(t,\hbar)$ satisfy the differential equation*

$$\left[D^{n+1-r} - e^t \prod_{j=1}^{r} l_j \prod_{m=1}^{l_j-1} (l_j D + \hbar m) \right] F(t,\hbar) = 0,$$

with $D = \hbar\frac{d}{dt} + l_1!\ldots l_r!e^t$.

Proof. Let f be a function of t. Then it is easy to prove by induction on m that

$$[D^m f(t)] e^{\frac{l_1!\ldots l_r!e^t}{\hbar}} = \left(\hbar\frac{d}{dt} \right)^m \left[f(t)e^{\frac{l_1!\ldots l_r!e^t}{\hbar}} \right].$$

This implies that

$$\left(\hbar\frac{d}{dt} \right)^m \left[s_\beta(t,\hbar)e^{\frac{l_1!\ldots l_r!e^t}{\hbar}} \right] = [D^m s_\beta(t,\hbar)] e^{\frac{l_1!\ldots l_r!e^t}{\hbar}}.$$

Moreover, if we develop the differential operator

$$e^t \prod_{j=1}^{r} l_j \prod_{m=1}^{l_j-1} \left(l_j\hbar\frac{d}{dt} + \hbar m \right)$$

as a polynomial in the powers of $\hbar\frac{d}{dt}$, we also obtain

$$\left\{ e^t \prod_{j=1}^{r} l_j \prod_{m=1}^{l_j-1} (l_j \hbar \tfrac{d}{dt} + \hbar m) \right\} \left[s_\beta(t,\hbar) e^{\frac{l_1!\ldots l_r! e^t}{\hbar}} \right] =$$

$$= \left\{ \left\{ e^t \prod_{j=1}^{r} l_j \prod_{m=1}^{l_j-1} (l_j D + \hbar m) \right\} [s_\beta(t,\hbar)] \right\} e^{\frac{l_1!\ldots l_r! e^t}{\hbar}}.$$

So the result follows.

□

Corollary 4.19. *In the small quantum cohomology algebra of X, $\dim X \neg 2$, the class p of hyperplane sections satisfies the following relation:*

$$(p + l_1! \ldots l_r! q)^{n+1-r} = l_1^{l_1} \ldots l_r^{l_r} q (p + l_1! \ldots l_r! q)^{n-r}.$$

Proof. It follows directly from Proposition 4.18 and Proposition 3.18

□

Example 4.20. *Let X be a non singular cubic surface in $\mathbb{P}^3$. Then the relation found in Corollary 4.19 is*

$$(p + 6q)^3 = 27q(6q + p)^2,$$

or

$$p^3 = 9qp^2 + 216q^2 p + 756q^3.$$

If we denote by $\{T_0 = 1, T_1 = p, p^2\}$ a basis for $H^*(X)$, then the intersection matrix is given by

$$\begin{pmatrix} 0 & 0 & 1 \\ 0 & 3 & 0 \\ 1 & 0 & 0 \end{pmatrix}.$$

In $SQH^*(X)$, we have

$$T_1 * T_1 = \Phi_{110} T_2 + \frac{\Phi_{111}}{3} T_1 + \Phi_{112} T_0$$

and

$$T_1 * T_1 * T_1 = \Phi_{110}(T_2 * T_1) + \frac{\Phi_{111}}{3}(T_1 * T_1) + \Phi_{112} T_1,$$

with Φ the potential involving Gromov-Witten invariants. Since

$$\Phi_{111} = qI_1(T_1, T_1, T_1) + \sum_{d \geq 2} q^d I_d(T_1, T_1, T_1),$$

and $T_1 * T_1 * T_1 = p^3$, this implies

$$9q(T_1 * T_1) = \frac{q}{3} I_1(T_1, T_1, T_1)(T_1 * T_1),$$

namely $I_1(T_1, T_1, T_1) = 27$. In other words, by the very definiton of Gromov-Witten invariants, there are 27 lines on a non-singular cubic surface in $\mathbb{P}^3$!

In order to prove Theorem 4.16, we proceed as in section 4. More precisely, we will determine explicitly the functions $Z_i(q, \hbar)$ defined in the previous section. In this case, we have

Proposition 4.21.

$$Z_i(q, \hbar) = \prod_{j \neq i} (\lambda_i - \lambda_j) \sum_{d \geq 0} \frac{q^d}{\hbar^d} \int_{\overline{\mathcal{M}}_{0,2}(\mathbb{P}^n, d)} \frac{e_0^*(\phi_i) \mathcal{E}'_{2,d}(-c)^{d-1}}{1 + \frac{c}{\hbar}},$$

with $q = e^{dt}$.

Proof. See Proposition 4.9 and keep in mind that $l = n$.

$\square$

In this case the recursive relations are more complicated because of the following

Lemma 4.22. Let $\Sigma \in \overline{\mathcal{M}}_{0,2}(\mathbb{P}^n, d)^{T^{n+1}}$ be a connected component of fixed points $[C, x_0, x_1; f]$ whose marked point x_0 is situated on a component of C with two or more special points. Then Σ gives zero contribution to the computation of the equivariant integrals

$$\int_{\overline{\mathcal{M}}_{0,2}(\mathbb{P}^n, d)} (-c)^k \mathcal{E}'_{2,d} e_0^*(\phi_i), \ k \geq d - 1,$$

unless $k = d$ and $C = C' \cup C''$, where C' is mapped to a fixed point p_i in $\mathbb{P}^n$ and carries both marked points, and C'' is a disjoint union of d irreducible components (intersecting C' at d special points) mapped (each with multiplicity 1) onto straight lines outgoing the point p_i (see figure 6 for $d = 5$).

Proof. By Proposition 4.10, the only non zero contribution comes from those components whose associated graph Γ (see section 4 for notations) satisfies the following requirements:

- there exists a vertex v with at least the label $\{0\}$;
- there are l_v edges issuing from v such that $h_v + l_v - 3 = d - 1$.

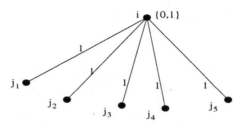

FIGURE 6.

This necessarily implies that $h_v = 2$ and $l_v = d$. Then the Lemma follows.

□

Set $z_i(Q, \hbar) = Z_i(\hbar Q, \hbar)$. We now compute the contribution of these connected components.

Proposition 4.23. *The non-zero contribution to*

$$\prod_{j \neq i} (\lambda_i - \lambda_j) \sum_{d \geq 0} Q^d \int_{\overline{\mathcal{M}}_{0,2}(\mathbb{P}^n, d)} (-c)^{d-1} \mathcal{E}'_{2,d} e_0^*(\phi_i)$$

given by connected components $\mathcal{K}$ described in Lemma 4.22 is

$$\exp \left\{ Q \frac{\prod_{a=1}^{r} (l_a \lambda_i - \mu_a)^{l_a}}{\prod_{\alpha \neq i} (\lambda_i - \lambda_\alpha)} \right\} \exp \left\{ -Q l_1! \ldots l_r! \right\}.$$

Proof. Let Γ be the graph associated with $\mathcal{K}$. By our assumptions, Γ is the graph (see figure 6 with $d = 5$) with $j_s \neq i$, $s \in \{1, \ldots, d\}$. Once we fix a $d-$ tuple $\{j_1, \ldots, j_d\}$, we denote by $\mathcal{K}_{j_1, \ldots, j_d}$ the corresponding connected component and by $\Gamma_{j_1, \ldots, j_d}$ its associatd graph. Next recall that by $\mathcal{N}_{j_1, \ldots, j_d}$ we mean the normal bundle $\mathcal{N}_{\overline{\mathcal{M}}_{0,2}(\mathbb{P}^n, d)/\mathcal{K}_{j_1, \ldots, j_d}}$. Let us consider the contribution to

$$\prod_{j \neq i} (\lambda_i - \lambda_j) \sum_{d \geq 0} Q^d \int_{\overline{\mathcal{M}}_{0,2}(\mathbb{P}^n, d)} (-c)^{d-1} \mathcal{E}'_{2,d} e_0^*(\phi_i)$$

given by all these connected components, i.e.

$$\prod_{j \neq i} (\lambda_i - \lambda_j) \sum_{d \geq 0} Q^d \sum_{(j_1, \ldots, j_d)} \int_{\mathcal{K}_{j_1, \ldots, j_d}} \frac{(-c)^{d-1} \mathcal{E}'_{2,d} e_0^*(\phi_i)}{\mathcal{E}(\mathcal{N}_{j_1, \ldots, j_d})}$$

where the second sum ranges over all d-tuples of integers j_s, $0 \leq j_s \leq n$, none of which equals i. Consider now the morphism

$$\delta := \delta_{j_1,\dots,j_d} : \mathcal{K}_{j_1,\dots,j_d} \to \overline{\mathcal{M}}_{0,2}(\mathbb{P}^n, d).$$

We are going to make explicit computations for the case $r = 1$ and then write down formulas in the general case. As in **STEP 1)** of Lemma 4.11, the localization of $e_0^*(\phi_i)$ at $\mathcal{K}_{j_1,\dots,j_d}$ is 1.

Since the component C'' (where x_0 is situated) of each point $[C, x_0, x_1; f] \in \mathcal{K}_{j_1,\dots,j_d}$ is mapped to the fixed point p_i in $\mathbb{P}^n$, T^{n+1} acts trivially on $\delta^* T^{(0)}$. By the localization formula, this means that $c_1(\delta^* T^{(0)})$ (c in the integral)$\in H^*(\mathcal{K}_{j_1,\dots,j_d})$. By Proposition 4.1 $\mathcal{K}_{j_1,\dots,j_d}$ is isomorphic to $\overline{\mathcal{M}}_{0,d+2}$ via $\xi_{\Gamma_{j_1,\dots,j_d}}$ and

$$\xi_{\Gamma_{j_1,\dots,j_d}}^* \left(c_1(\delta^* T^{(0)}) \right) = \tilde{c}_0,$$ where $\tilde{c}_0$ is the Chern class of *the universal tangent line bundle* of the marked point corresponding to x_0 on $\overline{\mathcal{M}}_{0,d+2}$.

We next compute the localization of $\mathcal{E}'_{2,d}$ at $\mathcal{K}_{j_1,\dots,j_d}$, i.e. the Euler class of $\delta^* W'_{2,d}$. By assumptions, we have $C = C_0 \cup C_1 \cup \dots C_d$, where each C_s is mapped to l_{ij_s} with degree one. Besides, denote by $q_s = C_0 \cap C_s$, $1 \leq s \leq d$. Since C_0 is mapped to the fixed point p_i, $\delta^* W'_{2,d}$ is trivial and thus its equivariant Euler class belongs to $H^*_{T^{n+1} \times T'}(B(T^{n+1} \times T'))$. If $z_{1,s}$ and $z_{2,s}$, $s = 1, \dots, d$, are homogeneous coordinates on C_s, then the action of T^{n+1} on C^s is

$$[z_{1,s}, z_{2,s}] \longrightarrow [t_i z_{1,s}, t_j z_{j,s}].$$

Therefore the character of the action of $T^{n+1} \times T'$ on a basis of $H^0(C_s, f_s^* \mathcal{O}(l))$, f_s being the restriction of f to C_s, is obtained as follows:

$$z_{1,s}^{l-m} z_{2,s}^m \to t_i^{l-m} t_j^m t'^{l-1} z_{1,s}^{l-m} z_{2,s}^m,$$

with $t' \in T'$ and $0 \leq m \leq l$. Hence the character is

$$m\lambda_{j_s} - \mu + (l - m)\lambda_i = (l\lambda_i - \mu) + m(\lambda_{j_s} - \lambda_i), \quad 0 \leq m \leq l.$$

By the definition of $W'_{2,d}$, the exact sequence holds

$$0 \to H^0(C, f^*\mathcal{O}(l)) \to \oplus_{s=1}^d H^0(C_s, f_s^*(\mathcal{O}(l))) \to \oplus_{s=1}^d f_s^*(\mathcal{O}(l)) \to 0 .$$
$$(\sigma_1, \dots, \sigma_d) \qquad (\sigma_1(q_1), \dots, \sigma_d(q_d))$$

Then the localization of $\mathcal{E}'_{2,d}$ at $\mathcal{K}_{j_1,\dots,j_d}$ is

$$\prod_{s=1}^d \prod_{m=1}^l [(l\lambda_i - \mu) + m(\lambda_{j_s} - \lambda_i)].$$

In general, for $r > 1$, we have

$$\prod_{a=1}^r \prod_{s=1}^d \prod_{m=1}^{l_a} [(l_a\lambda_i - \mu_a) + m(\lambda_{j_s} - \lambda_i)].$$

We now compute the localization of $\mathcal{E}(\mathcal{N}_{j_1,\ldots,j_d})$ at $\mathcal{K}_{j_1,\ldots,j_d}$. To this end, we use the observations made in Remark 3.9. Since the deformations which preserve the combinatorial type of C_0 and C_s, $1 \le s \le d$, are trivial, we need to compute contributions of those deformations induced by holomorphic vector fields on C. First of all, notice that the following exact sequence holds:

$$0 \to H^0(C, f^*T_{\mathbb{P}^n}) \to H^0(C_0, f_0^*T_{\mathbb{P}^n}) \oplus \oplus_{s=1}^d H^0(C_s, f_s^*T_{\mathbb{P}^n}) \to \oplus_{s=1}^d f^*T_{p_i} \to 0,$$

with $f_0 = f_{|C_0}$ and $f_s = f_{|C_s}$. For C_0, the only contribution to the normal bundle comes from $H^0(C_0, f_0^*T_{\mathbb{P}^n})$ which is a trivial vector bundle on $\mathcal{K}_{j_1,\ldots,j_d}$. Moreover, its equivariant Euler class is $\prod_{j \neq i}(\lambda_i - \lambda_j)$, since a basis is given by $f_0^*\left(\frac{\partial}{\partial X_j}\right)$, $j \in \left\{0,\ldots,\widehat{i},\ldots,n\right\}$.

The only contribution to the normal bundle stemming from deformations of each C_s is given by vector fields in

$$\mathcal{V} := \oplus_{s=1}^d \frac{H^0(C_s, f_s^*(T_{\mathbb{P}^n} - p_i))}{H^0(C_s, T_{C_s}(-q_s))}.$$

After observing (see **STEP 4**) of Lemma 4.12) that a basis for $\mathcal{V}$ is given by vector fields $z_1 f_s^*\left(\frac{\partial}{\partial X_\alpha}\right)$, $\alpha \in \left\{0,\ldots,\widehat{i},\widehat{j_s},\ldots,n\right\}$, $\mathcal{E}(\mathcal{V}) = \prod_{s=1}^d \prod_{\substack{\alpha \neq j_s \\ \alpha \neq i}}(\lambda_{j_s} - \lambda_\alpha)$.

Apart from these deformations, there is only the contribution of deformations smoothing the nodes. Thus, we can write

$$\mathcal{E}(\mathcal{N}_{j_1,\ldots,j_d}) = \mathcal{E}(\oplus_{s=1}^d (T_{q_s}C_0 \otimes T_{q_s}C_s)) \prod_{j \neq i}(\lambda_i - \lambda_j) \prod_{s=1}^d \prod_{\substack{\alpha \neq j_s \\ \alpha \neq i}}(\lambda_{j_s} - \lambda_\alpha).$$

Setting $\mathcal{E}(T_{q_s}C_s) = c_s$, then

$$\mathcal{E}(\oplus_{s=1}^d (T_{q_s}C_0 \otimes T_{q_s}C_s)) = \prod_{s=1}^d (c_s + \lambda_i - \lambda_{j_s}),$$

(see **STEP 4**) of Lemma 4.12 for the computation $\mathcal{E}(T_{q_s}C_s)$. We can also see this class as the pull back under $\xi_{\Gamma_{j_1,\ldots,j_d}}$ of the Chern class $\tilde{c}_s$, $1 \le s \le d$, of the *universal tangent bundle* of the marked point corresponding to q_s on $\overline{M}_{0,d+2}$.

To summarize, we obtain

$$\prod_{j \neq i}(\lambda_i - \lambda_j) \sum_{d \ge 0} Q^d \sum_{(j_1,\ldots,j_d)} \int_{\mathcal{K}_{j_1,\ldots,j_d}} \frac{(-c)^{d-1}\mathcal{E}_{2,d}' e_0^*(\phi_i)}{\mathcal{E}(\mathcal{N}_{j_1,\ldots,j_d})} =$$

$$= \sum_{d \ge 0} Q^d \sum_{(j_1,\ldots,j_d)} \frac{\prod_{a=1}^r \prod_{s=1}^d \prod_{m=1}^{l_a}[(l_a\lambda_i - \mu_a) + m(\lambda_{j_s} - \lambda_i)]}{\prod_{s=1}^d \prod_{\substack{\alpha \neq j_s \\ \alpha \neq i}}(\lambda_{j_s} - \lambda_\alpha)}.$$

$$\cdot \int_{\mathcal{K}_{j_1,\ldots,j_d}} \frac{(-c)^{d-1}}{\prod_{s=1}^d (c_s + \lambda_i - \lambda_{j_s})} \cdot$$

If we develop the cohomology class in the integral, we further have

$$\sum_{d \geq 0} Q^d \sum_{(j_1,\ldots,j_d)} \frac{\prod_{a=1}^r \prod_{s=1}^d \prod_{m=1}^{l_a}[(l_a\lambda_i - \mu_a) + m(\lambda_{j_s} - \lambda_i)]}{\prod_{s=1}^d \prod_{\substack{\alpha \neq j_s \\ \alpha \neq i}}(\lambda_{j_s} - \lambda_\alpha)} \cdot$$

$$\cdot \prod_{s=1}^d \sum_{k_s \geq 0} \frac{1}{(\lambda_i - \lambda_{j_s})^{k_s+1}} \int_{\mathcal{K}_{j_1,\ldots,j_d}} (-c)^{d-1}(-c_s)^{k_s+1} =$$

$$= \sum_{d \geq 0} Q^d \sum_{(j_1,\ldots,j_d)} \frac{\prod_{a=1}^r \prod_{s=1}^d \prod_{m=1}^{l_a}[(l_a\lambda_i - \mu_a) + m(\lambda_{j_s} - \lambda_i)]}{\prod_{s=1}^d \prod_{\substack{\alpha \neq j_s \\ \alpha \neq i}}(\lambda_{j_s} - \lambda_\alpha)} \cdot$$

$$\cdot \prod_{s=1}^d \sum_{k_s \geq 0} \frac{1}{(\lambda_i - \lambda_{j_s})^{k_s+1}} \int_{\overline{\mathcal{M}}_{0,d+2}} (-\tilde{c}_0)^{d-1}(-\tilde{c}_s)^{k_s+1}.$$

On the other hand, $(-\tilde{c}_i)$, $0 \leq i \leq d+2$, are Chern classes of *universal cotangent line bundles* of marked points on $\overline{\mathcal{M}}_{0,d+2}$. By dimension computations, it is clear that we have to compute the integral

$$\int_{\overline{\mathcal{M}}_{0,d+2}} (-\tilde{c}_0)^{d-1}$$

which, by standard results on intersection theory on $\overline{\mathcal{M}}_{0,d+2}$ is $\langle \tau_{d-1}, \tau_0, \ldots, \tau_0 \rangle = \langle \tau_0, \tau_0, \tau_0 \rangle = 1$ because of the string equation (see [W]).

In other words, we showed that

$$\prod_{j \neq i}(\lambda_i - \lambda_j) \sum_{d \geq 0} Q^d \sum_{(j_1,\ldots,j_d)} \int_{\mathcal{K}_{j_1,\ldots,j_d}} \frac{(-c)^{d-1}\mathcal{E}'_{2,d}e_0^*(\phi_i)}{\mathcal{E}(\mathcal{N}_{j_1,\ldots,j_d})} =$$

$$= \sum_{d \geq 0} Q^d(-1)^d \frac{1}{d!} \frac{\prod_{a=1}^r \prod_{s=1}^d \prod_{m=1}^{l_a}[(l_a\lambda_i - \mu_a) + m(\lambda_{j_s} - \lambda_i)]}{\prod_{s=1}^d \prod_{\substack{\alpha \neq j_s \\ \alpha \neq i}}(\lambda_{j_s} - \lambda_\alpha) \prod_{s=1}^d (\lambda_{j_s} - \lambda_i)} =$$

$$= \sum_{d \geq 0} Q^d(-1)^d \frac{1}{d!} \left(\sum_{j \neq i} \frac{\prod_{a=1}^r \prod_{m=1}^{l_a}[(l_a\lambda_i - \mu_a) + m(p - \lambda_i)]}{\prod_{\alpha \neq j}(p - \lambda_\alpha)} \right)^d \Bigg|_{p=\lambda_j} =$$

$$= \exp\left\{-Q\sum_{j\neq i}\frac{\prod_{a=1}^{r}\prod_{m=1}^{l_a}[(l_a\lambda_i - \mu_a) + m(p - \lambda_i)]}{\prod_{\alpha\neq j}(p - \lambda_\alpha)}\Bigg|_{p=\lambda_j}\right\}.$$

Let us now consider the function

$$\frac{\prod_{a=1}^{r}\prod_{m=1}^{l_a}[(l_a p - \mu_a) + m(p - \lambda_i)]}{\prod_{\alpha\neq j}(p - \lambda_\alpha)}$$

as a meromorphic function of p. Then by taking residues in $p = \lambda_\alpha$, $\alpha \neq j$, $p = \infty$ and $p = \lambda_i$, we have

$$\sum_{j\neq i}\frac{\prod_{a=1}^{r}\prod_{m=1}^{l_a}[(l_a p - \mu_a) + m(p - \lambda_i)]}{\prod_{\alpha\neq j}(p - \lambda_\alpha)}\Bigg|_{p=\lambda_j} = \frac{\prod_{a=1}^{r}(l_a\lambda_i - \mu_a)^{l_a}}{\prod_{\alpha\neq i}(\lambda_i - \lambda_\alpha)} + l_1!\ldots l_r!.$$

This completely proves the Lemma.

□

Theorem 4.24. *The functions $z_i(Q, \hbar)$ satisfy the recursive relations*

$$z_i(Q, \hbar) = 1 + \sum_{d>0}coeff_i(d)Q^d + \sum_{j\neq i}\sum_{d'>0}Q^{d'}coeff_i^j(d')z_j(Q, \frac{\lambda_j - \lambda_i}{d'}), \tag{4.6}$$

where

$$coeff_i^j(d') = \frac{\prod_{a=1}^{r}\prod_{m=1}^{l_a d'}[\frac{l_a\lambda' - \mu_a}{\lambda_j - \lambda_i}d' + m]}{d'\left(1 + \frac{\lambda_i - \lambda_j}{d'\hbar}\right)\prod_{\substack{\alpha=0 \\ (\alpha,m)\neq(j,d')}}^{n}\prod_{m=1}^{d'}[\frac{\lambda_i - \lambda_\alpha}{\lambda_j - \lambda_i}d' + m]},$$

and

$$coeff_i(d) = \sum_{t=1}^{d}\frac{1}{t!(d-t)!}\frac{\prod_{a=1}^{r}(l_a\lambda_i - \mu_a)^{l_a t}}{\prod_{\alpha\neq i}(\lambda_i - \lambda_\alpha)^t}(-l_1!\ldots l_r!)^{d-t}.$$

Proof. By Proposition 4.23 and Proposition 4.13, we obtain the recursive relations.

□

Corollary 4.25. *The functions* $z_i(Q, 1/\omega)$, $\omega = 1/\hbar$, *are power series* $\sum\limits_{d \geq 0} C_i(d, 1/\omega)Q^d$ *where the coefficients* $C_i(d, 1/\omega)$ *are rational functions of* ω *of the form*

$$C_i(d, 1/\omega) = \frac{P_{nd}^{(i)}\omega^{nd} + \ldots + P_0^{(i)}}{\prod\limits_{\alpha=0}^{m}\prod\limits_{m=1}^{d}[(\lambda_i - \lambda_j)\omega + m]},$$

$$C_i(0, 1/\omega) = 1.$$

The functions $z_i(Q, \hbar)$ *are uniquely determined by these properties, the recursive relations of Theorem 4.24 and the initial conditions*

$$\sum_{d>0} coeff_i(d)Q^d = \sum_{d>0} \frac{P_{nd}^{(i)}}{d! \prod\limits_{\alpha \neq i}(\lambda_i - \lambda_\alpha)^d}Q^d.$$

Proof. By the expression for $coeff_i(d)$ and $coeff_i^j(d')$, we find out that the coefficients $C_i(d, 1/\omega)$ have the desired form, since $d\sum_{a=1}^{r} l_a = dn$. Notice that the term $P_{nd}^{(i)}\omega^{nd}$ in the numerator of $C_i(d, 1/\omega)$ comes from $\sum\limits_{d>0} coeff_i(d)Q^d$. Conversely, the requirements of this Corollary completely determine the functions $z_i(Q, \hbar)$, since by 4.6, we have

$$C_i(0, 1/\omega) = 1,$$

$$C_i(d, 1/\omega) = coeff_i(d) + \sum_{j \neq i}\sum_{m=1}^{d} coeff_i^j(m)C_i(d - m, \frac{\lambda_j - \lambda_i}{m}).$$

$\square$

Theorem 4.26. *The series*

$$z_i(Q, 1/\omega) = \sum_{d \geq 0} Q^d \frac{\prod\limits_{a=1}^{r}\prod\limits_{m=1}^{l_a d}[l_a\lambda_i - \mu_a)\omega + m]}{d! \prod\limits_{\alpha \neq i}^{n}\prod\limits_{m=1}^{d}[(\lambda_i - \lambda_\alpha)\omega + m]} \exp\{-Ql_1! \ldots l_r!\}$$

satisfy the requirements of Corollary 4.25 and the initial conditions

$$1 + \sum_{d>0} coeff_i(d)Q^d = \exp\left\{Q\frac{\prod\limits_{a=1}^{r}(l_a\lambda_i - \mu_a)^{l_a}}{\prod\limits_{\alpha \neq i}(\lambda_i - \lambda_\alpha)}\right\}\exp\{-Ql_1! \ldots l_r!\}.$$

Proof. By developing the series $z_i(Q, \hbar)$, the coefficients $C_i(d, 1/\omega)$ have the form

$$\sum_{t=1}^{d} \frac{1}{t!(d-t)!} \frac{\prod\limits_{a=1}^{r} \prod\limits_{m=1}^{l_a t} [(l_a \lambda_i - \mu_a)\omega + m]}{\prod\limits_{\alpha \neq i}^{n} \prod\limits_{m=1}^{t} [(\lambda_i - \lambda_\alpha)\omega + m]} (-l_1! \ldots l_r!)^{d-t},$$

$(C_i(0, 1/\omega) = 1)$. Obviously, these coefficients have the desired form and the initial conditions are satisfied. In order to verify the recursive relations, set $z_i'(Q, \hbar) := z_i(Q, \hbar) \exp\{-l_1! \ldots l_r! Q\}$, and proceed similarly to Proposition 4.15.

□

Proof of Theorem 4.16 See Theorem 4.6.

□

4.4. THE CALABI-YAU CASE

In this section we are going to show how to relate solutions $s^*(t, \hbar)$ of the Picard-Fuchs equation for the mirror symmetric family of a Calabi-Yau projective complete intersection X in $\mathbb{P}^n$ and solutions of the equation $\nabla_\hbar s(T, \hbar) = 0$.

Once again, the two generating functions are different, and again this is due to the fact that Proposition 4.10 does not hold. Even in this case we can find recursive relations, even though of a much more complicated type, satisfied by coefficients of both functions, but with different initial conditions.

Givental makes use of certain "polynomial properties" to characterize a class $\mathcal{P}$ of solutions of these relations. These properties are satisfied by solutions of $\nabla_\hbar s(T, \hbar) = 0$ for *geometric reasons* (see Lemma 4.27 and Corollary 4.28) , and by solutions of the Picard-Fuchs for *computational reasons* (see Proposition 4.33).

Once this is established, Givental introduces a set of transformations that preserves the class $\mathcal{P}$, and describes precisely the effect of any such transformation on initial conditions; so it becomes clear how to recover functions $s^*(t, \hbar)$ from solutions $s_\beta(T, \hbar)$ of the equation $\nabla_\hbar s(T, \hbar) = 0$.

4.4.1. Properties and motivations arising from geometry.
We recall that

$$s_\beta(T, \hbar) = \int_{\mathbb{P}^n} P^\beta S_{(2)}(T, \hbar),$$

with $S_{(2)}(T, \hbar) = e^{\frac{PT}{\hbar}} \sum_{d \geq 0} e^{dT}(e_0)_* \left(\frac{\mathcal{E}_{2,d}}{\hbar + c} \right) \in H^*(\mathbb{P}^n)$.

Take now the equivariant counterpart $S'_{(2)}(T, \hbar)$ of $S_{(2)}(T, \hbar)$, and consider the function

$$\Phi(t, \tau) := \int_{\mathbb{P}^n}^{T^{n+1} \times T^r} \frac{S'_{(2)}(t, \hbar) S'_{(2)}(\tau, -\hbar)}{Euler(\oplus_i \mathcal{O}(l_i))}.$$

This function is exactly the $(0,0)$ entry of the matrix

$$\left\{ -\hbar^2 \frac{\partial^2 \mathcal{F}^1(x)}{\partial t_i \partial \tau_j} \right\}$$

described in section 3.2, with normal integrals replaced by equivariant ones.
From Theorem 3.13, taking in account that our basis $\{\phi_i\}$ is orthogonal, but not orthonormal, we can prove that

$$\Phi(t, \tau) = \sum_{d_1, d_2 \geq 0} e^{d_1 t} e^{d_2 \tau} \sum_{i=0}^n \frac{\prod_{a=1}^r (l_a \lambda_i - \mu_a)}{\prod_{j \neq i}(\lambda_i - \lambda_j)} \int_{\overline{\mathcal{M}}_{0,2}(\mathbb{P}^n, d_1)} \frac{e^{\left(\frac{Pt}{\hbar} \right)} \mathcal{E}'_{2,d_1}(e_0)^*(\phi_i)}{\hbar + c} \cdot$$
$$\cdot \int_{\overline{\mathcal{M}}_{0,2}(\mathbb{P}^n, d_2)} \frac{e^{\left(-\frac{P\tau}{\hbar} \right)} \mathcal{E}'_{2,d_2}(e_0)^*(\phi_i)}{-\hbar + c}.$$

Let

$$L_d := \left\{ \begin{array}{c} [C, x_0, x_1, (\psi_1, \psi_2)] \in \overline{\mathcal{M}}_{0,2}\left(\mathbb{P}^n \times \mathbb{P}^1, (d, 1) \right) : \\ \psi_2(x_0) \in \mathbb{P}^n \times \{0\}, \psi_2(x_1) \in \mathbb{P}^n \times \{\infty\} \end{array} \right\} \supset \left(\overline{\mathcal{M}}_{0,2}\left(\mathbb{P}^n \times \mathbb{P}^1, (d, 1) \right) \right)^{S^1}$$

and $L'_d := \mathbb{P}\left(\mathbb{C}^{n+1} \otimes H^0\left(\mathbb{P}^1, \mathcal{O}(d) \right) \right)$; points in this space are $(n+1)$-tuples of homogeneous polynomials of degree d in two variables.
The group $T = S^1 \times (S^1)^{n+1}$ acts on both spaces: the action on L_d has been already described, and the action on L'_d is induced by the diagonal action of $(S^1)^{n+1}$ on $\mathbb{C}^{n+1}$, and by rotations on $\mathbb{P}^1$; more precisely we consider the following action:

$$\begin{array}{ccc} S^1 & \times & \mathbb{P}^1 & \to & \mathbb{P}^1 \\ t & , & [z_1, z_2] & & [t^{-1} z_1, z_2] \end{array} \cdot \qquad (4.7)$$

Let $p = c_1^T(\mathcal{O}(1)) \in H_T^*(L'_d)$.

Lemma 4.27. "The Main Lemma". *There exists a regular T-equivariant map*

$$\mu : L_d \to L'_d,$$

such that $\Phi(t, \tau) = \sum_{d \geq 0} \exp(d\tau) \int_{L_d} \exp\left(\frac{\mu^*(p)(t - \tau)}{\hbar} \right) \mathcal{E}_{2,d}.$

First of all, let us give some motivation for proving this lemma; the proof is postponed at the end of this section.

Consider the bundle $F_{2,d}$ on L'_d such that substitution of the $n+1$ polynomials into r homogeneous equations of degrees $l_1, ..., l_r$ produces a section of this bundle. Introducing the addictional action of a torus $T' = (S^1)^r$ on this bundle, we can see that

$$H^*_{T \times T'}(L'_d) \cong \frac{\mathbb{Q}[p, \lambda_0, ..., \lambda_n, \mu_1, ..., \mu_r, \hbar]}{\prod_{j=0}^{n} \prod_{m=0}^{d} (p - \lambda_j - m\hbar)},$$

and that

$$\mathcal{F}_{2,d} := Euler^{T \times T'}(F_{2,d}) = \prod_{a=1}^{r} \prod_{m=0}^{l_a d} (l_a p - \mu_a - m\hbar).$$

Assume for a moment that $\mu^*(\mathcal{F}_{2,d}) = \mathcal{E}_{2,d}$; then substituting in the result of the lemma we wolud have that

$$
\begin{aligned}
\Phi(t, \tau) &= \sum_{d \geq 0} \exp(d\tau) \int_{L_d} \exp\left(\frac{\mu^*(p)(t - \tau)}{\hbar}\right) \mu^*(\mathcal{F}_{2,d}) = \\
&= \sum_{d \geq 0} \exp(d\tau) \int_{L'_d} \exp\left(\frac{p(t - \tau)}{\hbar}\right) \mathcal{F}_{2,d} = \\
&= \frac{1}{2\pi\sqrt{-1}} \cdot \\
&\quad \cdot \int \exp\left(\frac{p(t - \tau)}{\hbar}\right) \left(\sum_{d \geq 0} \exp(d\tau) \frac{\prod_{a=1}^{r} \prod_{m=0}^{l_a d} (l_a p - \mu_a - m\hbar)}{\prod_{j=0}^{m} \prod_{m=0}^{d} (p - \lambda_j - m\hbar)}\right) dp,
\end{aligned}
$$

and we would proceed exactly as in the simpler case $\sum_i l_i < n$. Unfortunately, the assumption is false, and consequently the function $\Phi(t, \tau)$ does not have exactly this form; anyway, the close relation between the spaces L_d and L'_d established in the lemma allows us to write a similar expression for $\Phi(t, \tau)$; namely, setting $q = \exp(\tau)$, $z = \frac{t - \tau}{\hbar}$, and $\Phi'(q, z) = \Phi(t, \tau)$, we can prove:

Corollary 4.28.

$$\Phi'(q, z) = \frac{1}{2\pi\sqrt{-1}} \int \exp(pz) \left(\sum_{d \geq 0} \frac{q^d E_d(p, \lambda, \mu, \hbar)}{\prod_{j=0}^{m} \prod_{m=0}^{d} (p - \lambda_j - m\hbar)}\right) dp,$$

where $E_d(p, \lambda, \mu, \hbar) = \mu_*(\mathcal{E}_{2,d})$ depends polynomially on all its variables.

Proof. Apply integration formula:

$$
\begin{aligned}
\Phi'(q, z) &= \sum_{d \geq 0} q^d \int_{L_d} \exp\left(\mu^*\left(p\right) z\right) \mathcal{E}_{2,d} = \\
&= \sum_{d \geq 0} q^d \int_{L'_d} \exp\left(pz\right) \mu_*\left(\mathcal{E}_{2,d}\right) = \\
&= \frac{1}{2\pi\sqrt{-1}} \int \exp.(pz) \left(\sum_{d \geq 0} \frac{q^d E_d\left(p, \lambda, \mu, \hbar\right)}{\prod_{j=0}^{m} \prod_{m=0}^{d}\left(p - \lambda_j - m\hbar\right)} \right) dp.
\end{aligned}
$$

Recall that $\mathcal{E}_{2,d} \in H^*_T(L_d)$ considered as a $\mathbb{Q}[\lambda_0, \ldots \lambda_n]$ module, hence a polynomial.

$\square$

Take now the equivariant counterpart $S'_{(2)}(T, \hbar)$ of $S(T, \hbar)$: if we compute the components of $S'_{(2)}(T, \hbar)$ with respect to the basis ϕ_i of $H_{T^{n+1} \times T^r}(\mathbb{P}^n)$ (see Lemma 4.8), the functions

$$
Z_i(q, \hbar) := \sum_{d \geq 0} q^d \int_{\overline{\mathcal{M}}_{0,2}(\mathbb{P}^n, d)} (e_0)^*(\phi_i) \frac{\mathcal{E}'_{2,d}}{\hbar + c}
$$

are not yet written in an explicit form.

4.4.2. The generating function for solutions of the Picard-Fuchs. Let us now define the function

$$
S^*_{(2)}(t, \hbar) = e^{\frac{Pt}{\hbar}} \sum_{d \geq 0} e^{dt} \frac{\prod_{a=1}^{r} \prod_{m=0}^{l_a d}(l_a P + m\hbar)}{\prod_{m=1}^{d}(P + m\hbar)^{n+1}}.
$$

On the basis of the analysis made in previous sections, we know that its components $s^*_i(t, \hbar)$ satisfy the Picard-Fuchs equation (see Corollary 4.3). We then consider the equivariant counterpart

$$
S'^*_{(2)}(t, \hbar) = e^{\frac{pt}{\hbar}} \sum_{d \geq 0} e^{dt} \frac{\prod_{a=1}^{r} \prod_{m=0}^{l_a d}(l_a p - \mu_a + m\hbar)}{\prod_{i=0}^{n} \prod_{m=1}^{d}(p - \lambda_i + m\hbar)^{n+1}}
$$

and introduce power series

$$
Z_i^*(q, \hbar) = \sum_{d \geq 0} q^d \frac{\prod_{a=1}^{r} \prod_{m=1}^{l_a d}(l_a \lambda_i - \mu_a + m\hbar)}{\prod_{\alpha=0}^{n} \prod_{m=1}^{d}(\lambda_i - \lambda_\alpha + m\hbar)} = \sum_{d \geq 0} q^d C_i^*(d, \hbar),
$$

so as to obtain, as proved in Lemma 4.8,

$$s_i'^*(t, \hbar) = \prod_{j \neq i} (\lambda_j - \lambda_i) e^{\frac{\lambda_i t}{\hbar}} \prod_{a=1}^{r} (l_a \lambda_i - \mu_a) Z_i^*(e^t, \hbar).$$

4.4.3. Recursive "Calabi-Yau" relations. First of all, we are going to prove that both $Z_i(q, \hbar)$ and $Z_i^*(q, \hbar)$ satisfy the same recursive relations. More precisely, as in subsections 4.2 and 4.3, we will determine recursive relations for $Z_i(q, \hbar)$, $0 \leq i \leq n$, with the aid of the integration formula over connected components of fixed points in $\overline{\mathcal{M}}_{0,2}(\mathbb{P}^n, d)$ and thereafter we will prove that $Z_i^*(q, \hbar)$ satisfy the same recursive relations.

Proposition 4.29.

$$Z_i(q, \hbar) = 1 + \sum_{d>0} \frac{q^d}{\hbar^d} \frac{R_{i,d}}{d!} + \sum_{d'>0} \sum_{j \neq i} \frac{q^{d'}}{\hbar^{d'}} \frac{coeff_i^j(d')}{\lambda_i - \lambda_j + d'\hbar} Z_j \left(\frac{q}{\hbar} \frac{\lambda_j - \lambda_i}{d'}, \frac{\lambda_j - \lambda_i}{d'} \right),$$

with $R_{i,d}$ a polynomial of $(\hbar, \lambda_\alpha, \mu_a)$ and

$$coeff_i^j(d') = \frac{\prod_{a=1}^{r} \prod_{m=1}^{l_a d'} \left[l_a \lambda_i - \mu_a + \frac{m}{d'}(\lambda_j - \lambda_i) \right]}{d'! \prod_{\alpha \neq i} \prod_{\substack{m=1 \\ (\alpha,m) \neq (j,d')}}^{d'} \left[\lambda_i - \lambda_\alpha + \frac{m}{d'}(\lambda_j - \lambda_i) \right]}.$$

Proof. If we expand $\frac{1}{\hbar+c}$ in power series, then

$$Z_i(q, \hbar) = 1 + \sum_{d>0} q^d \sum_{k \geq 0} \frac{1}{\hbar^{k+1}} \int_{\overline{\mathcal{M}}_{0,2}(\mathbb{P}^n, d)} \mathcal{E}_{2,d}' e_0^*(\phi_i)(-c)^k.$$

Since $\dim \overline{\mathcal{M}}_{0,2}(\mathbb{P}^n, d) = n + (n+1)d - 1$,

$$Z_i(q, \hbar) = 1 + \sum_{d>0} q^d \left[\sum_{k=0}^{d-1} \hbar^{-k-1} \int_{\overline{\mathcal{M}}_{0,2}(\mathbb{P}^n, d)} \mathcal{E}_{2,d}' e_0^*(\phi_i)(-c)^k \right] +$$

$$+ \sum_{d>0} q^d \hbar^{-d} \int_{\overline{\mathcal{M}}_{0,2}(\mathbb{P}^n, d)} \frac{\mathcal{E}_{2,d}' e_0^*(\phi_i)(-c)^d}{\hbar + c}.$$

By Proposition 4.10 and Lemma 4.22, the last sum is equal to

$$\sum_{d'>0} q^{d'} \hbar^{-d'} \frac{\prod_{a=1}^{r} \prod_{m=1}^{l_a d'} \left[l_a \lambda_i - \mu_a + \frac{m}{d'}(\lambda_j - \lambda_i) \right]}{d'! \prod_{\alpha \neq i} \prod_{\substack{m=1 \\ (\alpha,m) \neq (j,d')}}^{d'} \left[\lambda_i - \lambda_\alpha + \frac{m}{d'}(\lambda_j - \lambda_i) \right] (\lambda_j - \lambda_i + d'\hbar)} \cdot$$

$$\cdot Z_j \left(\frac{q}{\hbar} \frac{\lambda_j - \lambda_i}{d'}, \frac{\lambda_j - \lambda_i}{d'} \right).$$

On the other hand, the first sum gives the contribution $\frac{R_{i,d}}{d!}$. $\square$

Obviously, we have the following

Corollary 4.30. *The coefficients of the power series* $Z_i(q,\hbar) = \sum_{d\geq 0} q^d C_i(d,\hbar)$ *are rational functions*

$$C_i(d,\hbar) = \frac{P_d^{(i)}}{d!\hbar^d \prod_{j\neq i}\prod_{m=1}^d (\lambda_i - \lambda_j + m\hbar)},$$

where $P_d^{(i)}$ *is a polynomial in* $(\hbar, \lambda, \mu)$ *of degree* $(n+1)d$.

$\square$

Proposition 4.31. *The functions* $Z_i^*(q,\hbar)$ *satisfy the recursive relations of Proposition 4.29.*

Proof. See the proof of Proposition 4.15. Notice that initial conditions are obtained by dividing the numerator in $C_i^*(d,\hbar)$ by the denominator.

$\square$

In the sequel, recursive relations as the ones satisfied by $Z_i(q,\hbar)$, but possibly with $R_{i,d}$ polynomials of $(\hbar, \lambda_\alpha, \mu_\alpha)$ of degree at most d, will be referred to as "*Calabi -Yau relations*".
We can prove a result of uniqueness.

Proposition 4.32. *For any given polynomial* $R_{i,d}$, *the "Calabi -Yau relations" are satisfied by a unique solution of the form* $\sum_{d\geq 0}\left(\frac{q}{\hbar}\right)^d K_i(d,\hbar)$, *with* $K_i(0,\hbar) = 1$ *and* $K_i(d,\hbar)$ *rational functions of* $\hbar$ *as in Corollary 4.30 with numerator* $F_d^{(i)}$.

Proof. If we substitute functions $W_i(q,\hbar)$ in the relations, then we see that coefficients $K_i(d,\hbar)$ are uniquely determined by the relations

$$K_i(d,\hbar) = \frac{1}{d!}R_{i,d} + \sum_{j\neq i}\sum_{m=1}^d \frac{Coeff_i^j(m)}{m!(\lambda_j - \lambda_i + m\hbar)}\left(\frac{\lambda_j - \lambda_i}{m}\right)^{d-m} K_j\left(d-m, \frac{\lambda_j - \lambda_i}{m}\right).$$

$\square$

Both functions $Z_i(q,\hbar)$ and $Z_i^*(q,\hbar)$ enjoy another remarkable property.

4.4.4. **The class of solutions with polynomial properties.** Fix a generic solution $W_i(q,\hbar)$ of the "*Calabi-Yau relations*" and define the correlator

$$\Phi_{W_i}(q,z,\hbar) := \sum_{i=0}^n \frac{\prod_{a=1}^r (l_a\lambda_i - \mu_a)}{\prod_{j\neq i}(\lambda_i - \lambda_j)} e^{z\lambda_i} W_i(qe^{z\hbar},\hbar)W_i(q,-\hbar). \quad (4.8)$$

Let us compute the correlator for $Z_i(q,\hbar)$ and $Z_i^*(q,\hbar)$. To this end, we recall that L_d' denotes the projective space of $(n+1)$-tuples of polynomials in the complex variable z of degree at most d and p denotes the equivariant first Chern class of the Hopf bundle over L_d'.

Proposition 4.33.

$$\Phi_{Z_i^*}(q, z, \hbar) = \frac{1}{2\pi\sqrt{-1}} \int e^{pz} \sum_{d \geq 0} q^d \frac{\prod_{a=1}^{r} \prod_{m=0}^{l_a d}(l_a p - \mu_a + m\hbar)}{\prod_{\alpha=0}^{n} \prod_{m=0}^{d}(p - \lambda_\alpha - m\hbar)} dp =$$

$$= \int_{L_d'} e^{pz} \sum_{d \geq 0} q^d \prod_{a=1}^{r} \prod_{m=1}^{l_a d}(l_a p - \mu_a + m\hbar).$$

Proof. We have

$$\sum_{i=0}^{n} \frac{\prod_{a=1}^{r}(l_a \lambda_i - \mu_a)}{\prod_{j \neq i}(\lambda_i - \lambda_j)} e^{z\lambda_i} Z_i^*(qe^{z\hbar}, \hbar) Z_i^*(q, -\hbar) =$$

$$= \sum_{i=0}^{n} \frac{\prod_{a=1}^{r}(l_a \lambda_i - \mu_a)}{\prod_{j \neq i}(\lambda_i - \lambda_j)} e^{z\lambda_i} \left(\sum_{d_1 \geq 0} q^{d_1} e^{d_1 z\hbar} C_i^*(d_1, \hbar) \right) \left(\sum_{d_2 \geq 0} q^{d_2} C_i^*(d_2, -\hbar) \right) =$$

$$= \sum_{i=0}^{n} \frac{\prod_{a=1}^{r}(l_a \lambda_i - \mu_a)}{\prod_{j \neq i}(\lambda_i - \lambda_j)} e^{z\lambda_i} \sum_{d \geq 0} q^d \sum_{s=0}^{d} e^{sz\hbar} \left(\frac{\prod_{a=1}^{r} \prod_{m=1}^{l_a s}(l_a \lambda_i - \mu_a + m\hbar)}{\prod_{\alpha=0}^{n} \prod_{m=0}^{s}(\lambda_i - \lambda_\alpha + m\hbar)} \cdot \right.$$

$$\left. \cdot \frac{\prod_{a=1}^{r} \prod_{m=1}^{l_a(d-s)}(l_a \lambda_i - \mu_a - m\hbar)}{\prod_{\alpha=0}^{n} \prod_{m=0}^{(d-s)}(\lambda_i - \lambda_\alpha - m\hbar)} \right) =$$

$$= \sum_{i=0}^{n} \sum_{d \geq 0} q^d \sum_{s=0}^{d} e^{z(\lambda_i + s\hbar)} \frac{\prod_{a=1}^{r} \prod_{m=0}^{l_a d}(l_a \lambda_i - \mu_a - m\hbar + l_a s\hbar)}{\prod_{\substack{\alpha=0 \\ (\alpha,m) \neq (i,s)}}^{n} \prod_{m=0}^{d}(\lambda_i - \lambda_\alpha - m\hbar + s\hbar)} =$$

$$= \frac{1}{2\pi\sqrt{-1}} \int e^{pz} \sum_{d \geq 0} q^d \frac{\prod_{a=1}^{r} \prod_{m=0}^{l_a d}(l_a p - \mu_a + m\hbar)}{\prod_{\alpha=0}^{n} \prod_{m=0}^{d}(p - \lambda_\alpha - m\hbar)} dp.$$

$\square$

On the other hand, by the definition of $Z_i(q, \hbar)$, we have

$$\Phi_{Z_i}(q, z, \hbar) = \sum_{i=0}^{n} \frac{\prod_{a=1}^{r}(l_a \lambda_i - \mu_a)}{\prod_{j \neq i}(\lambda_i - \lambda_j)} e^{z\lambda_i} \cdot$$

$$\cdot \left(\sum_{d_1 \geq 0} q^{d_1} e^{d_1 z\hbar} \int_{\overline{\mathcal{M}}_{0,2}(\mathbb{P}^n, d_1)} \frac{\mathcal{E}_{2,d_1}'(e_0)^*(\phi_i)}{\hbar + c} \right) \cdot$$

$$\cdot \left(\sum_{d_2 \geq 0} q^{d_2} \int_{\overline{\mathcal{M}}_{0,2}(\mathbb{P}^n, d_2)} \frac{\mathcal{E}_{2,d_2}'(e_0)^*(\phi_i)}{-\hbar + c} \right).$$

Notice that, if we set $q = e^\tau$, $z = \frac{t-\tau}{\hbar}$, then $\Phi_{Z_i}(q, z, \hbar) = \Phi(t, \tau)$.

By the Main Lemma, we see that also for $Z_i(q, \hbar)$ the correlator $\Phi_{Z_i}(q, z, \hbar)$ can be written as an equivariant integral over L'_d, i.e.

$$\Phi_{Z_i}(q, z, \hbar) = \int_{L'_d} e^{pz} \sum_{d \geq 0} q^d E_d(p, \lambda_\alpha, \mu_a, \hbar),$$

with $E_d(p, \lambda_\alpha, \mu_a, \hbar)$ polynomials in p of degree at most $(n+1)d+n$ with polynomial coefficients in $(\lambda_\alpha, \mu_a, \hbar)$. We have pointed out two properties enjoyed by both $Z_i(q, \hbar)$ and $Z_i^*(q, \hbar)$. This leads us to define a particular class of solutions of the "Calabi-Yau relations".

Definition 4.34. *A solution $W_i(q, \hbar)$ of the "Calabi-Yau relations" belongs to $\mathcal{P}$ if $\Phi_{W_i}(q, z, \hbar)$ can be written as a sum of residues of polynomials $P_d(p, \lambda_\alpha, \mu_a, \hbar)$ in p of degree at most $(n+1)d + n$ with polynomial coefficients in $(\lambda_\alpha, \mu_a, \hbar)$.*

Notice that if $P_d(p, \lambda_\alpha, \mu_a, \hbar)$ exists, then it is uniquely determined. Indeed, the following holds.

Proposition 4.35. *Let $W_i(q, \hbar)$ be as in Proposition 4.32. The polynomial coefficients P_d in Φ_{W_i} are uniquely determined by their values*

$$P_d(\lambda_i + k\hbar) = \prod_{a=1}^r (l_a \lambda_i - \mu_a) F_k^{(i)}(\hbar) F_{d-k}^{(i)}(-\hbar), \ 0 \leq i \leq n, \ 0 \leq k \leq d.$$

Proof. If we compute the integral in Φ_{W_i} and compare with 4.8, then

$$\frac{P_d(\lambda_i + k\hbar)}{\prod_{\alpha=0}^n \prod_{\substack{s=0 \\ (\alpha,s) \neq (i,k)}}^d (\lambda_i - \lambda_\alpha + k\hbar - s\hbar)} = \frac{\prod_{a=1}^r (l_a \lambda_i - \mu_a) F_k^{(i)}(\hbar) F_{d-k}^{(i)}(-\hbar)}{\prod_{i \neq \alpha}(\lambda_i - \lambda_\alpha) k! \hbar^k \prod_{\alpha \neq i} \prod_{s=1}^k (\lambda_i - \lambda_\alpha + s\hbar)} \cdot$$

$$\cdot \frac{1}{(d-k)!(-\hbar)^{d-k} \prod_{i \neq \alpha} \prod_{s=1}^{d-k}(\lambda_i - \lambda_\alpha - s\hbar)} \cdot$$

Since

$$\prod_{\substack{\alpha=0 \\ (\alpha,s) \neq (i,k)}}^n \prod_{s=0}^d (\lambda_i - \lambda_\alpha + k\hbar - s\hbar) = \prod_{\alpha \neq i} \prod_{\substack{s=0 \\ s \neq k}}^d (\lambda_i - \lambda_\alpha + k\hbar - s\hbar) \prod_{\substack{s=0 \\ s \neq k}}^d (k-s)\hbar \prod_{\alpha \neq i}(\lambda_i - \lambda_\alpha) =$$

$$= \prod_{i \neq \alpha}(\lambda_i - \lambda_\alpha) k! \hbar^k \prod_{\alpha \neq i} \prod_{s=1}^k (\lambda_i - \lambda_\alpha + s\hbar)(d-k)!(-\hbar)^{d-k} \prod_{i \neq \alpha} \prod_{s=1}^{d-k}(\lambda_i - \lambda_\alpha - s\hbar),$$

the Proposition follows.

$\square$

It is possible to prove a result of uniqueness for functions in the class $\mathcal{P}$.

Proposition 4.36. *A solution from* $\mathcal{P}$ *is uniquely determined by the two first terms in the expansion of* $W_i(q, \hbar) = W_i^{(0)} + W_i^{(1)} \frac{1}{\hbar} + \ldots$ *as power series in* $\frac{1}{\hbar}$.

Proof. If we write $R_{i,d} = R_{i,d}^{(0)} \hbar^d + R_{i,d}^{(1)} \hbar^{d-1} + \ldots$, then

$$W_i^{(0)} = 1 + \sum_d R_{i,d}^{(0)} q^d,$$

$$W_i^{(1)} = \sum_d R_{i,d}^{(1)} q^d.$$

Fix $d \geq 0$ and suppose that $W_i(q, \hbar)$ and $W_i'(q, \hbar)$ are two solutions from $\mathcal{P}$ with $W_i^{(0)} = W_i^{'(0)}$ and $W_i^{(1)} = W_i^{'(1)}$. By induction it is possible to prove that $P_k = P_k'$ for each $k < d$ and thus P_d vanishes at $p = \lambda_i + k\hbar$, $0 \leq i \leq n$, $1 \leq k \leq d - 1$. This implies that $\prod_j \prod_{m=1}^{d-1} (p - \lambda_j - m\hbar)$ divides $P_d - P_d'$. In addition, we also have (see Proposition 4.35)

$$P_d(\lambda_i + d\hbar) - P_d'(\lambda_i + d\hbar) = \prod_{a=1}^{r} (l_a \lambda_i - \mu_a) F_d^{(i)}(\hbar) =$$

$$= \prod_{a=1}^{r} (l_a \lambda_i - \mu_a) d! \hbar^d \prod_{j \neq i} \prod_{m=1}^{d} (\lambda_i - \lambda_j - m\hbar) \{K_i(d, \hbar) - K_i'(d, \hbar)\} =$$

$$= \prod_{a=1}^{r} (l_a \lambda_i - \mu_a) \prod_{j \neq i} \prod_{m=1}^{d} (\lambda_i - \lambda_j - m\hbar) \left\{ R_{i,d} - R_{i,d}' \right\}.$$

So the polynomial $\left\{ R_{i,d} - R_{i,d}' \right\}$ should be divisible by $\hbar^{d-1}$ and, by our hypotheses, it is identically zero. Thus if two solutions in $\mathcal{P}$ coincide in orders $\hbar^0$ and $\hbar^{-1}$, then the very solutions coincide.

$\square$

4.4.5. Transformations in $\mathcal{P}$. At this point we come to describe transformations among solutions in $\mathcal{P}$.

Consider the following operations on the class $\mathcal{P}$:

(a) simultaneous multiplication $W_i(q, \hbar) \mapsto \nu_1(q) W_i(q, \hbar)$ by a power series of q with $\nu_1(0) = 1$;

(b) changes $W_i(q, \hbar) \mapsto e^{\frac{\lambda_i \nu_2(q)}{\hbar}} W_i(qe^{\nu_2(q)}, \hbar)$ with $\nu_1(0) = 0$;

(c) multiplication $W_i(q, \hbar) \mapsto \exp(C\nu_3(q)/\hbar) W_i(q, \hbar)$, where C is a linear function of (λ, μ) and $\nu_3(0) = 0$.

If we apply these operations to $W_i(q, \hbar)$, then we have to check that Φ_i transforms to an equivariant integral of power series of polynomials. To this purpose, we need two lemmas.

Lemma 4.37. *Suppose that a series*

$$s = \sum_d q^d P_d(p, \lambda_\alpha, \mu_a, \hbar)$$

with coefficients which are polynomials of p of degree $\leq \dim L'_d$, has the property that for every $k = 0, 1, \ldots$, the q series $\int_{L'_d} sp^k$ has polynomial coefficients in $(\lambda_\alpha, \mu_a, \hbar)$. Then the coefficients of all P_d are polynomials of $(\lambda_\alpha, \mu_a, \hbar)$ and vice versa.

Proof. Decomposing $P_d(p, \lambda_\alpha, \mu_a, \hbar) = \sum P_{dj}(\lambda_\alpha, \mu_a, \hbar)p^j$ we get that

$$\int_{L'_d} sp^k = \sum_d q^d \sum_j P_{dj}(\lambda_\alpha, \mu_a, \hbar)\frac{1}{2\pi\sqrt{-1}}\int \frac{p^{j+k}}{\prod_{\alpha=0}^n \prod_{m=0}^d (p - \lambda_\alpha - m\hbar)}dp;$$

the integral vanishes for $j+k < (d+1)(n+1)-1$, equals 1 for $j+k = (d+1)(n+1)-1$ and is a polynomial in $\lambda_\alpha, \hbar$ for $j+k > (d+1)(n+1)-1$. The resulting triangular matrix is invertible, and still the entries of the inverse matrix are polynomials in $\lambda_\alpha, \hbar$. Hence, on varying k, we obtain that every P_{dj} is a polynomial.

$\square$

Lemma 4.38. *The polynomiality of P_d in Φ_{W_i} is invariant with respect to the operations (a), (b), and (c).*

Proof. It is clear that multiplication by a series in q does not change the polynomiality property. If we perform a (b)-type operation and set $Q = qe^{\nu_2(q)}$, then

$$\sum_{i=0}^n \frac{\prod_{a=1}^r(l_a\lambda_i - \mu_a)}{\prod_{j\neq i}(\lambda_i - \lambda_j)}e^{z\lambda_i}W_i(Qe^{z\hbar - \nu_2(q)+\nu_2(qe^{z\hbar})}, \hbar)W_i(Q, -\hbar) =$$

$$= \int_{L'_d} e^{\frac{p}{\hbar}\{z\hbar - \nu_2(q)+\nu_2(qe^{z\hbar})\}}\sum_d q^d e^{d\nu_2(q)}P_d(p, \lambda_\alpha, \mu_a, \hbar).$$

By the previous Lemma, we have to prove that for all k the q series $\left(\frac{\partial}{\partial z}\right)^k_{|z=0}\Phi_{W_i}$ have polynomial coefficients. Since the exponent $z\hbar - \nu_2(q) + \nu_2(qe^{z\hbar})$ is divided by $\hbar$, the derivatives still have polynomial coefficients. The case of operation (c) is analogous with $\exp\left[C\frac{-\nu_3(q)+\nu_3(qe^{z\hbar})}{\hbar}\right]$.

$\square$

Theorem 4.39. *The class $\mathcal{P}$ is invariant with respect to the operations (a), (b), and (c).*

Proof. Consider the term $\frac{q^d}{\hbar^d}coef f_i^j(d)$ in the recursive relations. Application of the operations (a), (b), and (c) to the left- and right-hand side of the "*Calabi-Yau relations*" causes, respectively, the following modifications in this coefficient:

$$q^d \mapsto q^d \frac{f(q)}{f(Q)},$$

$$q^d \mapsto q^d \exp \left\{ \frac{\lambda_i f(Q\hbar)}{\hbar} - df(Q\hbar) - \frac{\lambda_i f(Q(\lambda_i - \lambda_j)/d)}{(\lambda_i - \lambda_j)/d} \right\},$$

$$q^d \mapsto q^d \exp \left\{ C\frac{f(Q\hbar)}{\hbar} - C\frac{f(Q(\lambda_j - \lambda_i)/d)}{(\lambda_j - \lambda_i)/d} \right\}.$$

In the case of the change (b), additionally, the argument $Q := (\lambda_j - \lambda_i)/d$ in Z_i, on the right-hand side of the recursive relation gets an extra-factor

$$\exp \left[f(Q\hbar) - f(Q(\lambda_j - \lambda_i)/d) \right]$$

. All the modifying factors above take 1 at $\hbar = (\lambda_j - \lambda_i)/d$. This means that the term responsible for the simple fraction with the pole at $\hbar = (\lambda_j - \lambda_i)/d$ does not change, and that the operations modify only the polynomial initial conditions. Under our assumptions about ν_1, ν_2, and ν_3, the modifying factors depend on $\hbar$ only in the combination $Q\hbar$. This implies that the degrees of the new initial conditions still do not exceed d.

The Theorem is completely proved by Lemma 4.38.

$\square$

4.4.6. Initial conditions. Let us determine initial conditions for $Z_i(q, \hbar)$ and $Z_i^*(q, \hbar)$.

Proposition 4.40. $Z_i^{(0)} = 1$, $Z_i^{(1)} = 0$.

Proof. The statements follows from the recursive relations, since the only term in degree zero is 1 and the contribution in degree 1 is given by the series

$$\sum_{d>0} q^d \int_{\overline{\mathcal{M}}_{0,2}(\mathbb{P}^n, d)} \mathcal{E}'_{2,d} e_0^*(\phi_i),$$

which is identically zero for dimension computations.

$\square$

Proposition 4.41.

$$Z_i^{*(0)} = \sum_{d \geq 0} \frac{(l_1 d)! \ldots (l_r d)!}{(d!)^{n+1}} q^d := f(q),$$

$$Z_i^{*(1)} = \lambda_i \sum_{a=1}^{r} l_a \left[g_{l_a}(q) - g_1(q) \right] + \left(\sum_{\alpha=0}^{n} \lambda_\alpha \right) g_1(q) - \sum_{a=1}^{r} \mu_a g_{l_a}(q),$$

where

$$g_l = \sum_{d \geq 1} q^d \frac{\prod_{a=1}^{r} (l_a d)!}{(d!)^{n+1}} \left(\sum_{m=1}^{ld} \frac{1}{m} \right).$$

Proof. If we set $\omega = 1/\hbar$, then

$$C_i^*(d,\omega) = \frac{\prod_{a=1}^r \prod_{m=1}^{l_a d} [(l_a\lambda_i - \mu_a)\omega + m]}{\prod_{\alpha=0}^n \prod_{m=1}^d [(\lambda_i - \lambda_\alpha)\omega + m]} = \frac{F(\omega)}{G(\omega)},$$

and so the statement follows easily.

$\square$

Finally, we can prove the following

Theorem 4.42. *Let $f(q)$ and $g_l(q)$ be as in Proposition 1.6. If we perform the following operations with $Z_i(q,\hbar)$:*

(1) put

$$Q = q\exp\left\{\sum_{a=1}^r l_a\left[g_{l_a}(q) - g_1(q)\right]/f(q)\right\},$$

(2) multiply $Z_i(Q(q),\hbar)$ by

$$\exp\left\{\frac{1}{f(q)\hbar}\left[\sum_{a=1}^r (l_a\lambda_i - \mu_a)g_{l_a}(q) - (\sum_{\alpha=0}^n (\lambda_i - \lambda_\alpha))g_1(q)\right]\right\},$$

(3) multiply all $Z_i(q,\hbar)$ simultaneously by $f(q)$.

Then the resulting functions coincide with functions $Z_i^*(q,\hbar)$.

Proof. The operations (1), (2) and (3) correspond to consecutive applications of operations of type (a), (b) and (c). Indeed, (3) is exactly (a) with $f(q) = \nu_1(q)$. In addition, if we set

$$\nu_2(q) = \left\{\sum_{a=1}^r l_a\left[g_{l_a}(q) - g_1(q)\right]/f(q)\right\},$$

then the change of variable in (1) is one of those allowed by operations of type (b). Finally, since $\sum_{a=1}^r l_a = n + 1$,

$$\frac{1}{f(q)\hbar}\left[\sum_{a=1}^r (l_a\lambda_i - \mu_a)g_{l_a}(q) - (\sum_{\alpha=0}^n (\lambda_i - \lambda_\alpha))g_1(q)\right] =$$

$$= \nu_2(q) - (\sum_{a=1}^r \mu_a)\frac{g_{l_a}(q)}{f(q)\hbar} + (\sum_{\alpha=0}^n \lambda_\alpha)\frac{g_1(q)}{f(q)\hbar},$$

and thus (2) corresponds to operations of type (c). (Notice that $f(q)$ is invertible). It is straightforward to verify that the initial conditions of $Z_i(q,\hbar)$ are transformed into those of $Z_i^*(q,\hbar)$. According to Proposition 1.10. this transforms functions $Z_i(q,\hbar)$ to $Z_i^*(q,\hbar)$.

□

4.4.7. Proof of the "Mirror Conjecture". The main and last theorem of this section is

Theorem 4.43. *Suppose $l_1 + \ldots + l_r = n + 1$. Consider the functions*

$$f(e^t) := \sum_{d \geq 0} \frac{(l_1 d)! \ldots (l_r d)!}{(d!)^{n+1}} e^{td}$$

and

$$g_s(e^t) := \sum_{d \geq 1} e^{td} \left(\sum_{m=1}^{sd} \frac{1}{m} \right) \frac{\prod_{a=1}^{r}(l_a d)!}{(d!)^{n+1}}.$$

After performing the change of variable

$$T = t + \sum_{a=1}^{r} \frac{l_a \left[g_{l_a}(e^t) - g_1(e^t) \right]}{f(e^t)},$$

and multiply the function $S_{(2)}(T, \hbar)$ by $f(e^t)$, we obtain the function

$$S_{(2)}^*(t, \hbar) = e^{\frac{Pt}{\hbar}} \sum_{d \geq 0} e^{dt} \frac{\prod_{a=1}^{r} \prod_{m=0}^{l_a d}(l_a P + m\hbar)}{\prod_{m=1}^{d}(P + m\hbar)^{n+1}}.$$

Proof. By Theorem 4.42, components of the equivariant counterpart of $S_{(2)}(T, \hbar)$, i.e.

$$s_i'(T, \hbar) = \prod_{j \neq i}(\lambda_j - \lambda_i) e^{\frac{\lambda_i T}{\hbar}} \prod_{a=1}^{r}(l_a \lambda_i - \mu_a) Z_i(e^T, \hbar),$$

can be transformed to components

$$s_i'^{*}(t, \hbar) = \prod_{j \neq i}(\lambda_j - \lambda_i) e^{\frac{\lambda_i t}{\hbar}} \prod_{a=1}^{r}(l_a \lambda_i - \mu_a) Z_i^*(e^t, \hbar).$$

If we pass to equivariant to non-equivariant cohomolgy, the Theorem follows.

□

Remark 4.44. *Notice that the components $s_0^*(t, \hbar)$ and $s_1^*(t, \hbar)$ in*

$$S_{(2)}^*(t, \hbar) = l_1 \ldots l_r \left[P^r s_0^*(t, \hbar) + P^{r+1} s_1^*(t, \hbar) + \ldots + P^n s_n^*(t, \hbar) \right]$$

are exactly $f(e^t)$ and $tf(e^t) + \sum_{a=1}^{r} l_a \left[g_{l_a}(e^t) - g_1(e^t) \right]$ respectively. Thus the inverse transformation from $S_{(2)}^*(t, \hbar)$ to $S_{(2)}(T, \hbar)$ consists in division by $f(e^t)$ followed by the change $T = \frac{s_1^*(t, \hbar)}{s_0^*(t, \hbar)}$ in accordance with the method conjectured by physicists [COGP].

4.4.8. Proof of the "Main Lemma". Let

$$L_d^0 := \overline{\mathcal{M}}_{0,0}\left(\mathbb{P}^n \times \mathbb{P}^1, (d,1)\right),$$

and let

$$\pi_0 : L_d \to L_d^0$$

be the map that forgets the two markings.
We shall construct a map

$$\mu_0 : L_d^0 \to L_d',$$

and then compose to define the map $\mu := \mu_0 \pi_0$.

STEP I) *Fixed points.*
We have already described fixed points in L_d^0 for the S^1 action (cfr. Prop. 3.8):

$$\left(L_d^0\right)^{S^1} = \bigcup_{k=0}^{d} M_{k,d-k}^0 =$$

$$= \bigcup_{k=0}^{d} \left\{ \begin{array}{l} [C_1 \cup C_0 \cup C_\infty, (\psi_1, \psi_2)] \in \left((Y \times \mathbb{P}^1)_{0,(d,1)}\right) : \\ \psi_{2|C_1} : C_1 \xrightarrow{\sim} \mathbb{P}^1, \psi_2(C_0) = 0, \psi_2(C_\infty) = \infty, \\ \deg \psi_{1|C_0} = k, \deg \psi_{1|C_\infty} = d - k, \psi_1(C_1) = \{pt.\} \end{array} \right\} \cong$$

$$\cong \bigcup_{k=0}^{d} Y_{1,k} \times_Y Y_{1,d-k}.$$

Among them, the fixed points for the $(S^1)^{n+1}$ action are given by

$$\bigcup_{i=0}^{n} \bigcup_{k=0}^{d} M_{k,d-k}^{0,i} \cong \bigcup_{i=0}^{n} \bigcup_{k=0}^{d} \nu^{-1}(p_i) \cap \left\{ (Y_{1,k})^{(S^1)^{n+1}} \times_Y (Y_{1,d-k})^{(S^1)^{n+1}} \right\},$$

where ν is the evaluation map on the marked point, and p_i is the projectivization of the i-th coordinate line in $\mathbb{P}^n$. Analogously,

$$\left(L_d\right)^{S^1} = \bigcup_{k=0}^{d} M_{k,d-k} \cong \bigcup_{k=0}^{d} Y_{2,k} \times_Y Y_{2,d-k},$$

and

$$\left(L_d\right)^{S^1 \times (S^1)^{n+1}} = \bigcup_{i=0}^{n} \bigcup_{k=0}^{d} M_{k,d-k}^i \cong$$

$$\cong \bigcup_{i=0}^{n} \bigcup_{k=0}^{d} \nu^{-1}(p_i) \cap \left\{ (Y_{2,k})^{(S^1)^{n+1}} \times_Y (Y_{2,d-k})^{(S^1)^{n+1}} \right\}.$$

On the other hand, the fixed points in L_d' are $(n+1)$-tuples of monomials of the following form:

$$\left[0, ..., 0, z_1^k z_2^{d-k}, 0, ..., 0\right], \ 0 \le k \le d.$$

STEP 2) *Definition and equivariancy of μ_0.*
There is a natural way to define μ_0 on the open set $\mathcal{M}_{0,0}\left(\mathbb{P}^n \times \mathbb{P}^1, (d,1)\right) \subset L_d^0$. In fact, a point in this set is an equivalence class of maps $\mathbb{P}^1 \to \mathbb{P}^n \times \mathbb{P}^1$ of bidegree $(d,1)$, and the image is the graph of a degree d map $\mathbb{P}^1 \to \mathbb{P}^n$, which is clearly given by a $(n+1)$-tuple of homogeneous polynomials of degree d in two variables, hence by a point in L_d'.
A generic point in L_d^0 is given by $[C_0 \cup C_1 \cup ... \cup C_r, (\psi_1, \psi_2)]$, with $\deg \psi_{2|C_0} = 1$, $\deg \psi_{2|C_i} = 0$, $i = 1,...,r$, and $\deg \psi_{1|C_j} =: d_j$, for every j, with $\sum d_j = d$.
Let $q_i := \psi_2(C_i)$, $i = 1,...,r$, and pick a homogenous polynomial g of degree $d' := d - d_0$, in two variables, with a root of order d_i in each q_i ; furthermore notice that the image of $\psi_{2|C_0}$ is the graph of a degree d_0 map $\mathbb{P}^1 \to \mathbb{P}^n$, hence is given by a $(n+1)$-tuple $[f_0, ..., f_n]$ of homogeneous polynomials of degree d_0.
   Now we are ready to define $\mu_0\left([C_0 \cup C_1 \cup ... \cup C_r, (\psi_1, \psi_2)]\right) := [gf_0, ..., gf_n]$; roughly speaking, the map μ_0 forgets everything about the components $C_1, ..., C_r$, except their image point in $\mathbb{P}^1$. Clearly,

$$\mu\left([C_0 \cup C_1 \cup ... \cup C_r, x_0, x_1, (\psi_1, \psi_2)]\right) = [gf_0, ..., gf_n].$$

By this definition, the map is $S^1 \times \left(S^1\right)^{n+1}$ equivariant; notice that

$$\mu\left(M_{k,d-k}^i\right) = \left[0, ..., 0, z_1^{d-k} z_2^k, 0, ..., 0\right].$$

STEP 2) *Regularity of μ_0.*
This is the most difficult step; we propose two different proofs of it; for both, we need to recall that the moduli space of maps can be also constructed ([FP]) as the GIT quotient of a quasi-projective variety $\mathcal{J}$, such that there exist a semi-universal family $\mathcal{F} \xrightarrow{\pi} \mathcal{J}$, and the fiber of this map on each point is isomorphic to the curve represented by the point itself.

   First proof.
We consider the following incidence variety:

$$\Gamma = \left\{ ([C, \psi_1], H, (p_1, ..., p_d)) \in \mathcal{J} \times (\mathbb{P}^n)^* \times \mathbb{P}^{1^{(d)}} : \psi_1^{-1}(H) \cap C = \psi_2^{-1}(p_1, ..., p_d) \right\},$$

where $\mathbb{P}^{1^{(d)}} \cong \mathbb{P}^d$ is the d-fold symmetric product of $\mathbb{P}^1$.
   Let

$$\mathcal{J} \times (\mathbb{P}^n)^* \times \mathbb{P}^{1^{(d)}} \xrightarrow{\gamma} \mathcal{J} \times (\mathbb{P}^n)^* \xrightarrow{\delta} \mathcal{J}$$

be the obvious projections. If we denote with $\Delta = \gamma(\Gamma)$, the map δ restricted to Δ is still surjective, since every degree d curve in $\mathbb{P}^n$ intersects a generic hyperplane exactly in d points; this allows us to say more, that is, the fiber $\delta^{-1}([C, \psi_1])$ in the following diagram

$$\Gamma \xrightarrow{\gamma} \Delta \xrightarrow{\delta} \mathcal{J}$$

is $\{[C, \psi_1]\} \times U_{[C,\psi_1]}$, where $U_{[C,\psi_1]}$ is the Zariski open subset of hyperplanes which intersect transversally the curve $\psi_1(C)$. Moreover, by definition of Γ, the fiber $\gamma^{-1}\delta^{-1}([C, \psi_1]) = \gamma^{-1}(\{[C, \psi_1]\} \times U_{[C,\psi_1]})$ can be viewed as $\{[C, \psi_1]\} \times G$, with G the graph in $U_{[C,\psi_1]} \times \mathbb{P}^{1^{(d)}}$ of a regular map $g_{[C,\psi_1]} : U_{[C,\psi_1]} \to \mathbb{P}^{1^{(d)}}$.

This map turns out to be linear, as it sends linear subspaces to linear subspaces, and we can extend it to a linear map $g_{[C,\psi_1]} : \mathbb{P}^n \to \mathbb{P}^d$, thus we can close up Γ to a subvariety $\overline{\Gamma}$ whose fiber over each curve is the graph of a linear map.

Consider the map

$$
\begin{array}{ccc}
\mathbb{P}^n \times \mathbb{P}\left(Hom\left(\mathbb{C}^{n+1}, \mathbb{C}^{d+1}\right)\right) & \overset{\alpha}{\to} & \mathbb{P}^n \times \mathbb{P}^d \\
[v], [f] & \to & [v], [f(v)]
\end{array}
$$

Our considerations imply that the subvariety $\widehat{\Gamma} := \left((Id, \alpha)^{-1}\left(\overline{\Gamma}\right)\right) \cap \mathcal{J} \times \{pt\} \times \mathbb{P}\left(Hom\left(\mathbb{C}^{n+1}, \mathbb{C}^{d+1}\right)\right)$ projects bijectively on the first factor $\mathcal{J}$; since this variety is normal, $\widehat{\Gamma}$ is isomorphic to the graph of a regular map

$$
\mathcal{J} \to \mathbb{P}\left(Hom\left(\mathbb{C}^{n+1}, \mathbb{C}^{d+1}\right)\right).
$$

Now we can easily observe that this map coincides set-theoretically with our μ_0, and we are done.

Second proof.

We are going to build a line bundle on $\mathcal{J}$ which naturally gives an invariant map $\mathcal{J} \to L'_d$, hence a map $L^0_d \to L'_d$, and then prove that coincides with our set-theoretically defined μ_0. We also need the map $e : \mathcal{F} \to \mathbb{P}^n \times \mathbb{P}^1$, the evaluation on the marked point. Finally, let $H := Hom\left(\mathcal{O}_{\mathbb{P}^n}(1), \mathcal{O}_{\mathbb{P}^1}(d)\right)$ be a line bundle on $\mathbb{P}^n \times \mathbb{P}^1$ (for simplicity, we omit to write the pullbacks by the projection maps). Consider the sheaf on $\mathcal{J}$ defined as follows:

$$
\mathcal{H}^0(U) := H^0\left(\pi^{-1}(U), e^*(H)\right).
$$

We claim:

1. $\mathcal{H}^0$ is a rank one locally free sheaf;
2. the fiber at $[C = C_0 \cup C_1 \cup ... \cup C_r, (\psi_1, \psi_2)]$ of the corresponding line bundle can be identified with $H^0\left(C_0, \psi^*_{|C_o}(H) \otimes \mathcal{O}(-q_1)^{d_1} \otimes ... \otimes \mathcal{O}(-q_r)^{d_r}\right)$.

We postpone the proof of these facts.

A non zero vector u_C in the above fiber gives a natural map

$$
f_C : H^0\left(C, \psi_1^*\left(\mathcal{O}_{\mathbb{P}^n}(1)\right)\right) \to H^0\left(C, \psi_2^*\left(\mathcal{O}_{\mathbb{P}^1}(d)\right)\right) = H^0\left(\mathbb{P}^1, \left(\mathcal{O}_{\mathbb{P}^1}(d)\right)\right);
$$

in fact, u_C is represented by a class $[\sigma] \in \mathcal{H}^0_{[C,\psi]}$; the restriction of σ to $\pi^{-1}([C, \psi])$ is well defined and does not depend on the choice of the element of the class; hence it is well defined

$$
[\sigma] \to \sigma_{|[C,\psi]} \in H^0\left(\pi^{-1}([C, \psi]), e^*(H)\right).
$$

Since $\pi^{-1}([C,\psi]) \cong C$, we can see $\sigma_{|\pi^{-1}([C,\psi])}$ as a section $\tau_C \in H^0(C, \psi^*(H))$
$= H^0(C, \psi^*(Hom(\mathcal{O}_{\mathbb{P}^n}(1), \mathcal{O}_{\mathbb{P}^1}(d))))$; the map f_C is constructed via τ_C.
Moreover, it will be clear after the proof of the claim that the kernel of the map f_C
consists exactly of sections vanishing identically on C_0; therefore we can proceed
as follows: pick a basis $\{X_0, ..., X_n\}$ of $H^0(\mathbb{P}^n, \mathcal{O}_{\mathbb{P}^n}(1))$, pull it back to a set of
elements in $H^0(C, \psi_1^*(\mathcal{O}_{\mathbb{P}^n}(1)))$, which cannot vanish simultaneously on C_0; the
map

$$\mathcal{H}^0_{[C,\psi]} \setminus \{0\} \rightarrow (\mathbb{C}^{n+1} \otimes H^0(\mathbb{P}^1, \mathcal{O}(d))) \setminus \{0\}$$
$$u_C \rightarrow f_C(\psi_1^*(X_0)), ..., f_C(\psi_1^*(X_n))$$

extends to a map

$$(\mathcal{H}^0)^* \rightarrow \mathbb{C}^{n+1} \otimes H^0(\mathbb{P}^1, \mathcal{O}(d)),$$

which is equivariant with respect to translation on the fiber on the LHS, and multiplication by a constant on the RHS. This map induces

$$\mathbb{P}(\mathcal{H}^0) \cong \mathcal{J} \rightarrow \mathbb{P}(\mathbb{C}^{n+1} \otimes H^0(\mathbb{P}^1, \mathcal{O}(d))) \cong L'_d,$$

which will be exactly our μ_0.

In order to justify claims 1 and 2, we need to compute the space of global sections
of the sheaf $e^*(H)$ in a formal neighborhood $\pi^{-1}(U)$ of the fiber $\pi^{-1}([C,\psi])$ of
the forgetful map $\pi : \mathcal{F} \rightarrow L^0_d$.
  A) *C is non singular.* In this case, we can find a formal neighborhood

$$\pi^{-1}(U) \cong C \times \Delta_\epsilon,$$

with $\Delta_\epsilon = Spec\mathbb{C}[[\epsilon]]$.
  We can prove that

$$\mathcal{H}^0(U) = H^0(\pi^{-1}(U), e^*(H)) \cong \mathbb{C}[[\epsilon]] \otimes H^0(C, \psi^*(H));$$

in fact, we can build an obvious covering of $\pi^{-1}(U)$ trivializing the bundle, namely
$\mathcal{W} = \{W_1, W_2\}$, where

$$W_i \cong \mathbb{C} \times \Delta_\epsilon$$

with coordinates (x_i, ϵ) related by $x_1 = x_2^{-1}$. Let $g_{12}(x_2, \epsilon)$ be the transition
function of the bundle $e^*(H)$ with respect to this covering; notice that

$$H^0(\pi^{-1}([C,\psi]), e^*(H)) \cong H^0(C, \psi^*(H)) = \mathcal{O},$$

hence $g_{12}(x_2, 0) = k$ and $g_{12}(x_2, \epsilon) = k(1 + \epsilon f_{12}(x_2, \epsilon))$, $k \in \mathbb{C}$, is an invertible
function .
A section in $H^0(\pi^{-1}(U), e^*(H))$ is given by $\{\sigma_1(x_1, \epsilon), \sigma_2(x_2, \epsilon)\}$ with

$$\sigma_1(x_1, \epsilon) = g_{12}(x_2, \epsilon)\sigma_2(x_2, \epsilon) = k(1 + \epsilon f_{12}(x_2, \epsilon))\sigma_2(x_2, \epsilon);$$

thus we can set $\tau_2(x_2, \epsilon) := (1 + \epsilon f_{12}(x_2, \epsilon))\sigma_2(x_2, \epsilon)$, and we can establish a
bijection between couples $\{\sigma_1, \sigma_2\}$ and couples $\{\sigma_1, \tau_2\}$.

By developing in power series with respect to ϵ, we see that the latter one is an element of $\mathbb{C}[[\epsilon]] \otimes H^0\left(C, \psi^*\left(H\right)\right) \cong \mathbb{C}[[\epsilon]]$. The map f_C is consequently a multiplication by a constant. In particular $f_C\left(\psi^*\left(X_i\right)\right) = kf_i$, as we required.

B) *C is reducible.* For the sake of simplicity, suppose there are two irreducible components, $C = C_0 \cup C_1$. Once more $\pi^{-1}\left([C, \psi]\right) \cong C$; if q is the node, we choose a formal neighborhood $\pi^{-1}\left(U\right)$ with the following trivializing covering for $e^*\left(H\right)$:

$$W \cong \Delta_{x_0, x_1, \epsilon}$$

is a neighborhood of q, and the curve has equation $x_0 x_1 = 0, \epsilon = 0$;

$$W_0 \cong \mathbb{C} \times \Delta_{y_0, \eta, \epsilon}$$

is a neighborhood of $C_0 \backslash \{q\}$, and the curve has equation $\eta = \epsilon = 0$, and

$$W_1 \cong \mathbb{C} \times \Delta_{y_1, \eta, \epsilon}$$

is a neighborhood of $C_1 \backslash \{q\}$, and the curve has equation $\eta = \epsilon = 0$; moreover, we have transition functions $y_0 = x_0^{-1}, \eta = x_0 x_1$ from W to W_0, and $y_1 = x_1^{-1}, \eta = x_0 x_1$ from W to W_1.

In order to see how $e^*\left(H\right)_{|\pi^{-1}\left([C, \psi]\right)}$ looks like, it is useful to observe that

$$\psi^*_{|C_0}\left(Hom\left(\mathcal{O}_{\mathbb{P}^n}\left(1\right), \mathcal{O}_{\mathbb{P}^1}\left(d\right)\right)\right) \cong \mathcal{O}\left(d - d_0\right),$$

and

$$\psi^*_{|C_1}\left(Hom\left(\mathcal{O}_{\mathbb{P}^n}\left(1\right), \mathcal{O}_{\mathbb{P}^1}\left(d\right)\right)\right) \cong \mathcal{O}\left(-d_1\right).$$

Let $\{\sigma\left(x_0, x_1, \epsilon\right), \sigma_0\left(y_0, \eta, \epsilon\right), \sigma_1\left(y_1, \eta, \epsilon\right)\}$ be a section in $H^0\left(\pi^{-1}\left(U\right), e^*\left(H\right)\right)$, with

$$\sigma\left(x_0, x_1, \epsilon\right) = g_1\left(y_1, \eta, \epsilon\right)\sigma_1\left(y_1, \eta, \epsilon\right);$$

since $\psi^*\left(H\right)_{|C_1} \cong \mathcal{O}\left(-d_1\right)$, $g_1\left(y_1, \eta, \epsilon\right) = y_1^{d_1} f_1\left(y_1, \eta, \epsilon\right)$, with $f_1\left(y_1, \eta, \epsilon\right)$ invertible. We obtain $\sigma\left(x_0, x_1, \epsilon\right) = y_1^{d_1} f_1\left(y_1, \eta, \epsilon\right)\sigma_1\left(y_1, \eta, \epsilon\right)$; observe that we should have $\sigma_1\left(y_1, \eta, \epsilon\right) = \eta^{d_1} \tau_1\left(y_1 \eta, \epsilon\right)$, otherwise, since $y_1 = x_1^{-1}$, σ could not be a polynomial of positive degree in x_1, hence $f_1^{-1}\sigma = y_1^{d_1} \eta^{d_1} \tau_1\left(y_1 \eta, \epsilon\right) = x_0^{d_1} \tau_1\left(x_0, \epsilon\right)$; this means that from $f_1^{-1}\sigma$, which is a function on W with a zero of order $\geq d_1$ in q, we recover σ_1.

Now we proceed as in **A)** with the couple $\{\sigma, \sigma_0\}$, related by a transition function $g_0\left(y_0, \eta, \epsilon\right) = y_0^{-d_0} f_0\left(y_0, \eta, \epsilon\right)$, and prove that there is a bijection between triples $\{\sigma, \sigma_0, \sigma_1\}$ which give a section in $H^0\left(\pi^{-1}\left(U\right), e^*\left(H\right)\right)$, and couples $\{\sigma, f_0\sigma_0\}$ which give rise to an element of

$$\mathbb{C}[[\epsilon]] \otimes H^0\left(C_0, \psi^*_{|C_0}\left(H\right) \otimes \mathcal{O}\left(-q\right)^{d_1}\right)$$

thus

$$\mathcal{H}^0\left(U\right) \cong \mathbb{C}[[\epsilon]] \otimes H^0\left(C_0, \psi^*_{|C_0}\left(H\right) \otimes \mathcal{O}\left(-q\right)^{d_1}\right).$$

Since $\dim H^0 \left(C, \psi^*_{|C_0} (H) \otimes \mathcal{O} (-q)^{d_1} \right) = 1$, this implies our claims in this case.

Moreover, the couple $\{\sigma, f_0\sigma_0\}$, evaluated in $\epsilon = 0$ should give a homogeneous polynomial of degree $d_1 = d - d_0$ on C_0, but we already proved that it has a zero of order d_1 in q, and this determines it completely; if g is this polynomial, then in this case the map f_C consists of multiplication by g, and $f_C (\psi^* (X_i)) = g f_i$; an induction procedure on the number of irreducible components completes the argument. Note that the map coincides on points which correspond to isomorphic stable maps, hence descends to a map on L_d^0, and that it is exactly μ_0.

STEP 3) Let us look at the expression of $\sum_{d \geq 0} \exp (d\tau) \int_{L_d} \exp \left(\frac{\mu^*(p)(t-\tau)}{\hbar} \right) \mathcal{E}_{2,d}$. The bijection established at the end of **STEP 1)** between connected components of fixed points in L_d and L_d' tells us that the localization of $\mu^*(p)$ at $M_{k,d-k}^i$ equals $\lambda_i + k\hbar$, and thus the pullback of $\mu^*(p)$ to the fixed point set

$$M_{k,d-k} := \bigcup_{i=0}^{n} M_{k,d-k}^i \cong \left\{ (Y_{2,k})^{(S^1)^{n+1}} \times_Y (Y_{2,d-k})^{(S^1)^{n+1}} \right\}$$

coincides with the $\left(S^1\right)^{n+1}$ equivariant class $\nu^* (p + k\hbar)$. Hence, by computations of section 3.2.2, taking in account that the action of S^1 is slightly different, and recalling conventions for degenerate cases, the following holds:

$$\sum_{d \geq 0} \exp (d\tau) \int_{L_d} \exp \left(\frac{\mu^* (p) (t - \tau)}{\hbar} \right) \mathcal{E}_{2,d} =$$

$$= \sum_{d \geq 0} \exp (d\tau) \sum_{k=0}^{d} \int_{M_{k,d-k}} \frac{\exp \left(\frac{\nu^*(p+kh)(t-\tau)}{\hbar} \right)}{\mathcal{N}_{M_{k,d-k}/L_d}} \mathcal{E}_{2,d} =$$

$$= \sum_{d \geq 0} \exp (d\tau) \sum_{k=0}^{d} \sum_{i=0}^{n} \frac{\prod_a (l_a \lambda_i - \mu_a)}{\prod_{j \neq i} (\lambda_j - \lambda_i)} \cdot$$

$$\cdot \left(\exp (kt) \int_{Y_{2,k}} \frac{e_0^* (\phi_i) \exp \left(\frac{pt}{\hbar} \right)}{\hbar + c} \mathcal{E}'_{2,d} \right) \cdot \left(\exp (-k\tau) \int_{Y_{2,d-k}} \frac{e_0^* (\phi_i) \exp \left(-\frac{p\tau}{\hbar} \right)}{-\hbar + c} \mathcal{E}'_{2,d-k} \right) =$$

$$= \sum_{i=0}^{n} \frac{\prod_a (l_a \lambda_i - \mu_a)}{\prod_{j \neq i} (\lambda_j - \lambda_i)} \sum_{d \geq 0} \sum_{k=0}^{d} \left(\exp (kt) \int_{Y_{2,k}} \frac{e_0^* (\phi_i) \exp \left(\frac{pt}{\hbar} \right)}{\hbar + c} \mathcal{E}'_{2,d} \right) \cdot$$

$$\cdot \left(\exp ((d - k) \tau) \int_{Y_{2,d-k}} \frac{e_0^* (\phi_i) \exp \left(-\frac{p\tau}{\hbar} \right)}{-\hbar + c} \mathcal{E}'_{2,d-k} \right) =$$

$$= \sum_{i=0}^{n} \frac{\prod_a (l_a \lambda_i - \mu_a)}{\prod_{j \neq i} (\lambda_j - \lambda_i)} \sum_{d,d' \geq 0} \exp(dt) \int_{Y_{2,k}} \frac{e_0^* (\phi_i) \exp\left(\frac{pt}{\hbar}\right)}{\hbar + c} \mathcal{E}'_{2,d}.$$

$$\cdot \exp(d'\tau) \int_{Y_{2,d'}} \frac{e_0^* (\phi_i) \exp\left(-\frac{p\tau}{\hbar}\right)}{-\hbar + c} \mathcal{E}'_{2,d'} = \Phi(t,\tau).$$

$\square$

REFERENCES

[AB] M. Atiyah and R. Bott, *The moment map and equivariant cohomology*, Topology 23 (1984), 1-28.

[Al] P. Aluffi, *Quantum cohomology at the Mittag-Leffler institute, 1996-1997*, Appunti della Scuola Normale Superiore, (1998).

[Be] A. Beauville, *Quantum cohomology of complete intersections*, Mathematical Physics, Analysis, Geometry 2 (1995), 384-398.

[BS] V. V. Batyrev and D. van Straten, *Generalized hypergeometric functions and rational curves on Calabi-Yau complete intersections in toric varieties*, Comm. Math. Phys. 168 (1995), 493-533.

[COGP] P. Candelas, X. C. de la Ossa, P. S. Green, and L. Parkes, *A pair of Calabi Yau manifolds as an exact soluble superconformal theory*, Nuc. Phys. B 359 (1991), 21-74.

[Du] B. Dubrovin, *Geometry of 2D topological field theories*, in *Integrable systems and quantum groups*, Montecatini Terme, 1993, Lecture Notes in Mathematics, 1620, Springer-Berlin, 1996, 120-348.

[FP] W. Fulton, R. Pandharipande, *Notes on the stable maps and Quantum Cohomology*, math.AG/9608011 (1996).

[G1] A. B. Givental, *Equivariant Gromov-Witten Invariants*, IMRN No.13 (1996), 613-663.

[G2] A. B. Givental, *Homological geometry I: Projective hypersurfaces*, Selecta Math I (1995), 325-345.

[G3] A. B. Givental, *Homological geometry and mirror symmetry*, in *Proceedings of the International Congress of Mathematicians, 1994, Zürich*, Birkhäuser, Basel, (1995), 472-480.

[Ji] M. Jinzenji, *On quantum cohomology rings for hypersurfaces in $\mathbb{CP}^{N-1}$*, hep-th 9511206.

[K] M. Kontsevich, *Enumeration of rational curves via torus actions*, in *The moduli space of Curves*, ed. by R. Dijkgraaf, C. Faber, and G. van der Geer, Progr Math. 129, Birkhäuser (1995), 335-368.

[LLY] B. H. Lian, K. Liu, S-H Yau, *Mirror Principle I*, math.AG/9712011 (1997).

[Ma] Y. Manin, *Generating functions in algebraic geometry and summation over trees*, in *The moduli space of Curves*, ed. by R. Dijkgraaf, C. Faber, and G. van der Geer, Progr Math. 129, Birkhäuser (1995), 401-417.

[Pa] R. Pandharipande, *Rational curves on hypersurfaces (after A. Givental)* , math.AG/9806133 (1998).

[Vo] C. Voisin, *Variations of Hodge structure of Calabi-Yau threefolds*, Quaderni della Scuola Normale Superiore (1998).

[W] E. Witten, *Two dimensional gravity and intersection theory on moduli space*, Surveys in Diff. Geom. I, (1991), 243-310.

SCUOLA NORMALE SUPERIORE, P.ZZA DEI CAVALIERI,7, 56126 PISA, ITALIA

DIPARTIMENTO DI MATEMATICA "ISTITUTO GUIDO CASTELNUOVO", UNIVERSITÀ DI ROMA "LA SAPIENZA", P.ZZALE A. MORO, 2, 00185 ROMA, ITALIA

E-mail address, Gilberto Bini: `bini@cibs.sns.it`
E-mail address, Corrado De Concini: `deconcin@mat.uniroma1.it`
E-mail address, Marzia Polito: `polito@cibs.sns.it`
E-mail address, Claudio Procesi: `claudio@mat.uniroma1.it`

Elenco dei volumi della collana
"Appunti"
pubblicati dall'Anno Accademico 1994/95

GIUSEPPE BERTIN (a cura di), *Seminario di Astrofisica*, 1995.

GIUSEPPE DA PRATO, *Introduction to Differential Stochastic Equations*, 1995.

EDOARDO VESENTINI, *Introduction to continuous semigroups*, 1996.

LUIGI AMBROSIO, *Corso introduttivo alla Teoria Geometrica della Misura ed alle Superfici Minime*, 1997.

CARLO PETRONIO, *A Theorem of Eliashberg and Thurston on Foliations and Contact Structures*, 1997.

MARIO TOSI, *Introduction to Statistical Mechanics and Thermodynamics*, 1997.

MARIO TOSI, *Introduction to the Theory of Many-Body Systems*, 1997.

PAOLO ALUFFI (a cura di), *Quantum cohomology at the Mittag-Leffler Institute*, 1997.

GILBERTO BINI, CORRADO DE CONCINI, MARZIA POLITO, CLAUDIO PROCESI, *On the Work of Givental Relative to Mirror Symmetry*, 1998

"CompoMat" Loc. Braccone, 02040 Configni (RI), Italy
Finito di stampare nell'agosto 1998